C016189210

DEATH ON EARTH

Also available in the Bloomsbury Sigma series:

Sex on Earth by Jules Howard
p53 – The Gene that Cracked the Cancer Code by Sue Armstrong
Atoms Under the Floorboards by Chris Woodford
Spirals in Time by Helen Scales
Chilled by Tom Jackson
A is for Arsenic by Kathryn Harkup
Breaking the Chains of Gravity by Amy Shira Teitel
Suspicious Minds by Rob Brotherton
Herding Hemingway's Cats by Kat Arney
Electronic Dreams by Tom Lean
Sorting the Beef from the Bull by Richard Evershed and Nicola Temple

DEATH ON EARTH

ADVENTURES IN EVOLUTION AND MORTALITY

Jules Howard

Bloomsbury Sigma
An imprint of Bloomsbury Publishing Plc

50 Bedford Square
London
WC1B 3DP
UK

1385 Broadway
New York
NY 10018
USA

www.bloomsbury.com

First published 2016

Photo credits (t = top, b = bottom, l = left, r = right, c = centre)

Colour section: P. 1: Ullstein bild / Getty Images. P. 2: Teresa L. Iglesias (A, B, D) and Gail L. Patricelli (C) (t); BSIP Ducloux/ Brisou/ Science Photo Library (cl); RedTC/ Shutterstock (bl). P. 3: Beverly Joubert / Getty Images (t); Auscape/ UIG/ Getty Images (b). P. 4: Steve Russell/ Contributor/ Getty Images (t); Thierry Falise/ Contributor/ Getty Images (b). P. 5: Jules Howard (tr); Katarina Christenson/ Shutterstock (tl); Tomatito/ Shutterstock (cl); Kurt_G/ Shutterstock (bl); Henrik Larsson/ Shutterstock (cr,br). P. 6: Chicago Tribune/ Getty Images (t); / Shutterstock (cl); CreativeNature, R. Zwerver/ Shutterstock (bl). P. 7: Vancouver Aquarium Marine Science Centre (t); Silviu Petrovan/ Froglife (www.froglife.org) (cr); Barcroft/ Contributor/ Getty Images. P. 8: Barcroft/ Getty Images (t); Erlendur Bogason (b); Bangor University (inset)

British Library Cataloguing-in-Publication Data
A catalogue record for this book is available from the British Library.

Library of Congress Cataloguing-in-Publication data has been applied for.

ISBN (hardback) 978-1-4729-1507-8
ISBN (trade paperback) 978-1-4729-1508-5
ISBN (ebook) 978-1-4729-1510-8

2 4 6 8 10 9 7 5 3 1

Illustrations by Samantha Goodlet

Typeset in Bembo Std by Deanta Global Publishing Services, Chennai, India
Printed and bound in Great Britain by CPI Group (UK) Ltd,
Croydon CR0 4YY

Bloomsbury Sigma, Book Twelve

For Lettie and Esme

Contents

INTRODUCTION

Neck ligaments. A cross-section of a trachea. An eyeball. In front of me is a white shelf filled with cylinders of varying sizes that contain a host of pickled parts. Swollen human hands, bleached spines and knee joints, some sort of sawn-off skull cap, brains in jars. This is not my usual day out. Some body parts are in cylinders, some in Perspex rectangular cubes; all are resting within some unknown embalming fluid that seems to bleach things in just the right nightmarish sort of way.

I walk to the next set of shelves. I stop. I sip my coffee and calmly put it back on its saucer. I realise I am shaking – the teaspoon on my plate starts knocking rhythmically against the coffee cup like I am a tiny alarm clock. I am ringing, and people start to look my way. I try to gather myself. The irony is that I am genuinely a bit alarmed by all of this. The room is about the size of two tennis courts – there is a good space in the middle, overlooked by two tiers of metal balconies that loom above us. The glass ceiling covers the hundred or so attendees to the event in a sepia glaze, as though we're in a bizarre lucid dream from which we can't escape. For a century or more this enormous room was an operating theatre; thousands of medical procedures and post-mortems have been undertaken here. And it really was a theatre: the light from above, the balconies in tiers that would once have been home to hundreds of students, eager to perfect their science and their future trade. This is a strange place to be.

I am attending something called Death Salon, an American movement holding its first event in the UK. It's taking place here, at Barts Pathology Museum, just around the corner from the financial sector of the City of London. According to the blurb on the welcome pack in my hand, Death Salon is an event 'that brings together intellectuals and independent thinkers engaged in the exploration of our shared mortality by sharing knowledge and art'. It had sounded really interesting, partly because I have always wanted to be an 'independent thinker'. I am pleased to report that I have achieved this aim admirably. Here I am independently thinking about things like tracheae, intestines and tiny testicles bobbing up and down in a preservative solution. I am here for three whole days, I realise. Three whole days.

I look around at the other attendees – there certainly is quite a mix of punters here, old and young. It is the first conference that I have attended for years where the gender bias is very female-heavy. This is a welcome change. I'm struck by how many young people there are, too – not just studious sorts, either. The style here isn't exactly preppie. It's something … it's something I've never seen before. Many of the attendees have a kind of … *mortician chic.* The men have so much style – geek glasses are in, as are leather satchels and skinny jeans. Some wear pinstripe suits and trainers. I notice one man wearing bowling shoes and managing, inexplicably, to pull it off. And the women, too. There's an air of burlesque about some: flowing curls, long clinging black dresses and black nail varnish. Many of them have fringes. Not for the first time in my life, I stick out like a sore thumb. I continue drinking my coffee, my little coffee spoon trilling like my beating heart.

My literary agent Jane was one of the people who had told me about Death Salon, and she's actually attending. I see her in the front row talking to another of her clients, a kind-looking lady who is here to give a presentation about her experiences of grief after the death of her mother. Just as the presentations are beginning I stumble down a row of chairs, heading to one of only a handful of empty ones at the far end. I catch Jane's eye from the back of the room. She gives me a thumbs-up and looks at me gleefully. She mouths the words 'ISN'T THIS WONDERFUL?' from across the room. I give her a slightly saggy thumbs-up back. I find a seat and gather myself. I'm here to write a book, Goddammit. I'm here to begin my journey.

Everyone says not to start work on a book until it's been commissioned. Only now do I realise why. Like Alice, I'm falling into a rabbit hole from which I'm struggling to escape – if the book isn't commissioned, this is a journey I will end up making unpaid and my family will, for a few years, hate me for it. Jane and I are still waiting for the green light but, what the hell, I'm starting anyway. Death, life and evolution seems like too interesting a topic to ignore any longer, and hopefully the book will be commissioned so everything is going to be ok, I think. Jim, my editor at Bloomsbury, has recently been giving me little supportive messages, but he's deeply worried that the whole death idea won't pan out. His normal friendly manner has become edgy of late; I can tell that he's worried that his colleagues won't go for it. One of Jim's concerns is that he doesn't think a book will sell if it has 'Death' in the title. But he has other worries …

One of which is that he doesn't think a book will sell if it has the word 'death' on most pages. Jim has warned me that people don't like to think about death. And that people don't like to buy books about people that think about death. Jim doesn't like it when people don't buy books, which is why he's advised throughout NOT TO START THIS BOOK until it's green-lit. He's anxious. That's ok, I tell him – this book will be different. I won't be writing a clichéd book about how DEATH IS NATURAL and that THERE'S NOTHING WE CAN DO ABOUT IT because people *always* say stuff like this and it's all got a little patronising.

In human terms – well, for me – I hate thinking about death. I hate it. I haven't written a will. I haven't got a retirement plan. I have no health insurance. Plus, I have just discovered that I also have a surprise suppressed squeamishness about human body parts in flasks of alcohol. But I *love* nature. I love evolution and the myriad ways in which natural selection makes and creates things more incredible even than humans can imagine. I love the diversity. The variety. The varieties. I love the colour, the roles, the niches, the behaviours, the true stories, the magic, the wonder. Surely death plugs into these wonders? I had said to Jim. Surely death is the universal thing that awaits all of these things? I had pitched the idea a few weeks ago to him. 'Jim,' I said. 'Jim, there is a story to tell about the impact that death has had on evolution, and on the niche-filled planet on which we find ourselves in the twenty-first century. I think there are miraculous acts that death imbibes into nature; acts that power it; acts that power its diversity. I want to chart this. I want to chart death's impact on nature and evolution and look at it in the context of our own mortality.'

I have unfinished business with death, after all. In my previous book *Sex on Earth* I had brought together a collection of ideas about sex in the animal kingdom. In the book I felt the world (and particularly some parts of the media) needed to appreciate a wider view of sex in nature, unburdened by human interests about whose penis is larger and which animal can orgasm for the longest. I stood up for pandas as creatures as fully evolved for sex as anything else. I took on penis-obsessed news editors, extolling the virtues of studying female reproductive anatomy alongside studies of male reproduction. I shouted up for mites and slugs and spiders and, I hope, allowed readers to re-evaluate their opinions about creatures many would rather step on than sexually appraise. I stood up for diversity. I stood up for sex. But with every story there was a nagging problem that never left me. It was simple: animals evolve to become masters of sex but … why don't they evolve to avoid death, or to live longer lives? This question was in my mind the whole time. Think about this for a second, if you will. Why must everything die? Why can't multicellular life evolve ways to replenish cells for longer, thereby allowing them greater opportunities for sex? Surely genes for such modifications would flourish, so why is it not something we see more of? Why is death so pervasive in nature? Why aren't there more immortals, whose genes could theoretically flood gene pools with sexual survivors? Questions, questions, questions. We live on a planet where life shares one primary drive: to make more of itself. After four billion years of evolution the world has filled up with animals that survive and reproduce ably. But death? Why would *that* persist? Why hasn't natural selection fixed death and filled the world up with immortals? It's not like death happens in one or two genera, or families on the periphery. *Everything dies*, I had thought. Somehow, it powers life and everything we see around us. Why? Why is the world like this? Questions, questions, questions.

I'm not totally new to the science of death, by any means. As I mentioned, I have had a deep interest in animal sex for many years, and there are a host of examples where the life principles of sex and death rub comfortably up against one another in such stories. Famous examples include those salmon species that migrate as juveniles from rivers into the ocean, and then return to rivers to spawn, where they then die. The female spiders (and possibly mantids) that devour their male partners during sex. The female mites that have evolved not to lay eggs externally, but instead allow their offspring to hatch from eggs within their body and then eat the female from the inside out. The female toads that often drown after being grabbed and wrestled by seven or eight eager males during breeding bouts. The mayfly species that live as larvae in freshwaters for a year or two yet live for only a matter of days as paid-up flying sexual adults. You'll know all of these stories, I'm sure. But these are just for starters. We all are the product of liaisons between creatures that got sex and death in the correct order. Untold trillions didn't, and untold trillions don't.

But there are other phenomena related to death that appear throughout the zoological literature which are simply a bit odd and make no immediate sense: tortoises that can survive for centuries; caterpillars that, according to some definitions of life and death, die – that become cellular goo within a chrysalis then manage, inexplicably, to reorganise into an animal we call a butterfly. And then there is the bigger picture: why is it that 99 per cent of species are already extinct? How does death contribute to life? What does it give us? And what causes cells to age? Can ageing be stopped? Can we live forever? And, would we really want to? This is where the scientific rubs up uncomfortably with the mortal mind and the modern experience of being an animal in a modern human world. I wondered if I could cross that line and try to understand why, on the whole, we humans are a little bit strange about death.

Researching the sex lives of animals for my previous book, I felt enormously appreciative of the scientists whom I interviewed. Each was making great strides in our understanding of the evolution of sex, and many were trying to explain exactly why it is so prevalent across the tree of life. They wanted to talk about it. They loved it. But death is equally prevalent across nature and I haven't ever heard anyone, really openly and clearly, explain much about it in zoological terms. It seemed to me to be a part of the biological sciences still kept in the dark – frowned upon, maybe. Ignored. Spooky, perhaps. I am drawn to topics like these, it seems ... So it seemed like an interesting one for a zoological writer to explore. We all know death, but less about the science. Nearly all of us will know the shock of death, the awfulness, the suffering and the deep life-changing impact that the death of close family and friends (and pets, of course) has on us. But do we talk about the zoological side of death? No, we don't. So let's give it a go, I said to Jim. Let's do it. And so my journey had started. I was beginning, even if I was still waiting on the book being commissioned.

Over the three days of Death Salon, something that started out as a slightly ghoulish and macabre experience developed into something totally different – it actually became a place of life. I listened to what living people are legally allowed to do with the dead bodies of their relatives. I heard about the history of CPR. I learned exactly how organ donation works, and how bodies can be donated to science. I saw graffiti-decorated coffins. I saw my first CT scan of an autopsy. I sat there, agog, as someone took off all of her clothes and we were asked to draw her holding a golden skull, her *memento mori*. I saw a virtual human autopsy. At one point a man stood up to tell us about a miniature railway he'd built with a toy-town cemetery, and how each tiny plastic ghost had been carefully teased off a job lot of novelty Halloween earrings he'd bought at Claire's Accessories. I saw people smiling and laughing and laughing – yes laughing, having fun – in the face of death. But over those three days at the conference I was always an observer; a bit of a loner who was sat quietly in the corner. Death Salon opened up my eyes to human death, yet no one, in the three days of the conference, mentioned biology. My world – of science, of life, of evolution – was barely mentioned. I found this rather strange. Death, as we know it, is a biological condition; it's that *other* whirring cog in natural selection's clock.

What follows in this book is my attempt at unravelling the many complicated threads surrounding biological death. The word 'journey' in popular science is pretty much the most overused descriptive going, and so I can only apologise about this. But this book really did become a journey – a journey, I hope, as life-affirming as it gets. It's a journey without taboos or (I hope) clichés. A journey guided by science at (almost) all times. A journey through the minds of the scientists that study it. And a study of the great merry(ish) journey that all life must take, from birth, to sex, to death and back in some other earthly form that is probably, at some point, going to be worm-like. And, predictably, it was a journey that almost killed me … Thank goodness (and Jim) it got commissioned. Thank goodness (and Jim) I got back from where it took me. Whether anyone chooses to read it is, of course, another story. But thanks, at least, for getting this far.

PART ONE

THIS IS A DEAD FROG

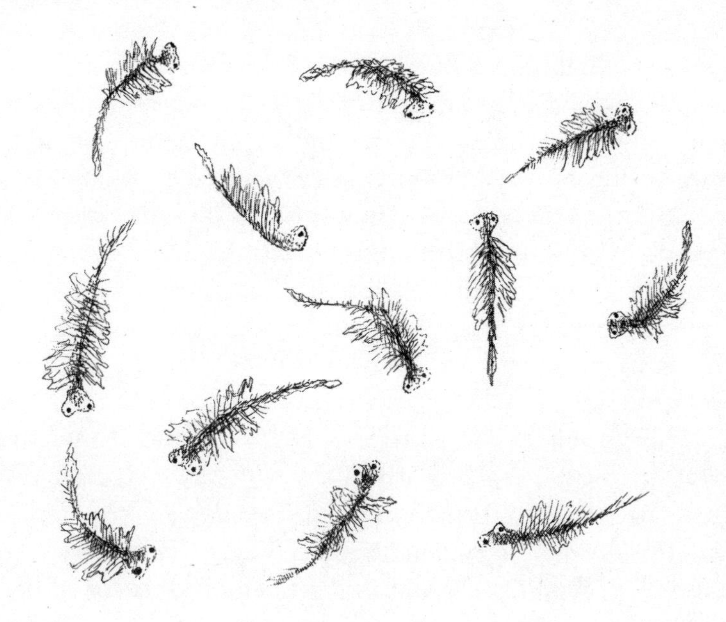

CHAPTER ONE

Life and Death in the Universe

What is life? thought Erwin Schrödinger. The answer was simple: it was something to fill his time. A subject to write a book about. So he did. Though many know him best for cats in boxes (or not), it's with life that his ideas first united and then fragmented domains of science. He published *What is Life?* in 1944. Based on a series of public lectures at Trinity College in Dublin the year before, Schrödinger's book really was a rare gem: a fairly readable account of the how and why and what of life on Earth. Four hundred people attended Schrödinger's original lectures on the subject, even though they came with a warning that 'the subject-matter was a difficult one … even though the physicist's most dreaded weapon, mathematical deduction, would hardly be utilized.' *What is Life?* really was breathtaking, though. Not only did it attempt to reconcile the world of biology with the realms of chemistry and physics, it was also an early contender for first speculating on the existence of an 'aperiodic crystal' that could carry genetic information through the generations in complex configurations of molecules (read: DNA). Among the many things covered in the book, Schrödinger identified life as perhaps the biggest paradox in the universe. Life just shouldn't occur, he realised. Yet it does. He attempted to explain how and why.

Think about the universe. It's chaotic. Mightily chaotic. Suns burn brightly – their energy radiating off into more loosely organised forms of energy (you and I call this, mostly, heat). Mountains erode. Continents split. Complex chemistry, produced under pressure or from lightning that befell planets like ours, falls apart with time. Radioactive particles smash and split. Chaos reigns. It really does. In the language of physicists, states drift naturally toward entropy – disorder, chaos, mixed-upness. Allow me to offer an analogy to explain this. The classic analogy of states drifting toward disorder involves libraries. Picture yours now … now imagine there were no librarians there. Imagine people coming and going from the library, taking books and bringing them back, every week or so. Some of the books will be put back in the wrong place or will be left haphazardly on tables or on top of the shelves. Give it a few weeks and you'd barely notice a big change, of course. Go back in a year, however, and you will probably start to have trouble finding what you want. There will be gaps in the shelves; some of the books will be stacked horizontally or left on the floor; the astrology books will be muddled up with the astronomy books; the spines of the romantic fiction will be worn; the Malcolm Gladwell books will have all their pages folded. You get the idea. Come back in two years and things will be even worse. Things will be messy. Five years, worse still. Ten years later? Total chaos. Come back in 200 years and the library will barely be standing. Come back in 500 and there will be nothing of the books but dust. Come back in a million years and the strata will contain nothing but the mud upon which the building stands. And that's if you're lucky. And that's chaos, ladies and gentlemen. Without someone (or something) being paid to maintain order, things drift into mixed-upness – that's how nearly all things work (it's also a good argument for why we should pay librarians much more than we currently do).

This understanding of states moving endlessly toward disorder (in a closed system) was first offered up by Newton: it was, famously, his Second Law of Thermodynamics. It explains everything we see out there. Except for in jellyfish. Or hamsters, for that matter. Or worms. Or walruses. Or wallflowers. Or winkles or white-tailed sea-eagles or, well, you get the idea. For in life, something strange happens. Cells don't leak and slop into one another after 10 minutes or 10 hours. Bodies don't just erode and fall apart and become functionless everywhere one looks. They are complex. And they remain so throughout life. Their patterns and make-up are the antithesis of disorder. Bodies are honed machines, and they work. They remain, throughout life, ordered. Unrotten. And this is rather strange when you think about it, because so few other things in the universe manage this.

'How does the living organism avoid decay?' This is the question Schrödinger attempted to tackle in *What is Life?* And the answer, he realised, is that it pays. To temporarily avoid death throughout their life, animals must pay. They must invest energy. In this respect, their cells and cell processes are like career librarians, holding back the chaos. Pushing against inevitability. Eventually the cost will become too high and they cease to be, of course – they die. Disorder whirls out from their organised bodies, much of which is recycled back into order within the life of others on Earth. But there is more even to life than this,

Schrödinger realised. Far from it being a freakish unexpected one-off in the universe and flying in the face of Newton's Second Law, physicists like Schrödinger realised that there was a certain inevitability to life. For there is a universal quirk of animals which many of us take for granted: we take energy from the sun (albeit by eating plants that have taken energy from the sun or animals that have eaten plants) and we produce heat. We make a highly disordered form of energy (heat) from an ordered one (light). We are part of the chaos, in other words – as well as paying the universe for the thrill of living, we also help maintain its universal tendency toward chaos. We fit right in. As absurd as it sounds, you and I are nothing more than heat pumps (though some pump more heat than others).

So, life plays by the rules. Life emerges because … well, because it can. It fits into the universe's way of behaving; it emerges because it fulfils a natural universal tendency toward an overall increase in entropy, essentially. Schrödinger's explanation was a good one. And like all good explanations, it has stuck around precisely because it has proved so hard an argument to better. One assumes that all life across the universe will obey this fundamental law; that life can be defined by its energetics. So that's it. All finished. We've defined life. That's that done then, right? Well, no. For there are other definitions of life, and there are other definitions of death. And it is with these definitions that my journey begins.

A grandfather stands with his five-year-old grandson in the busy canteen of one of the world's finest natural history museums. He faces a dilemma. The queue for the cafe is very long and it'll take him ages to get served.

He can't stand around with a grumpy five-year-old for 15 minutes, he thinks. But he needs a coffee, desperately. What should he do? He thinks. He considers his options. And then he makes a decision. He does something no one else does in modern times: he looks around the tables in the busy canteen for people who look friendly and caring and who definitely won't take his young grandchild off in a van if he asks to leave him with them. He scans the room. Who looks trustworthy? There are older people scattered in little groups talking to one another. Not them, he thinks. There is a small group of students. Not them either. A lady on her own reading a book? No, not her. And then he sees them. There are two people animatedly discussing something, talking very deeply and looking very intensely at one another while one of them scribbles things in a notebook. One – a woman – is friendly and talking passionately. The other – a man – is furrowing his brow and keeps looking up at the ceiling while scratching his chin pretentiously. Those people are us.

Perfect, the older gentleman thinks, for reasons I now struggle to comprehend. He walks over to us with his young grandchild in tow. 'Excuse me,' he says with slight reticence. 'Excuse me, but would you take my grandchild while I go and queue?' He rethinks the wording of his sentence. 'Will you … *look after* my grandchild while I get a coffee?' 'Oh …' we say. 'I'll be right over there,' he says. He points at the coffee bar. We have been chosen. We are trustworthy. It's quite a nice feeling. We smile a little nervously. 'Erm, yes … sure, ok …' says Louisa, smiling politely. 'I'm only round the corner,' says the gentleman. 'Just in the queue for a coffee … you know, it'll be fine.' He makes it sound like the most normal thing in the world to leave his grandchild with strangers. In some ways it is. Or it was. It's an honour, really; an honour to be chosen for looking like the people least likely to steal this child. 'Erm, yes, of course, yes,' I say. 'Yes, that's fine. Sure.' The boy sits down, looking a bit uncomfortable at having to sit with two complete strangers. 'Sit *straight* Harry,' says the grandfather. 'Come on, sit right.' He jiggles Harry about in his chair. The boy shuffles and sits up straight facing us both, still with his eyes locked firmly down at the floor. He looks up at us sheepishly and places his little hands on the table. The grandfather trundles off to join the back of a very long queue. We both look at the young lad. 'Hi,' I say. He says nothing. 'Hi,' says Louisa. He says nothing again. He doesn't know it yet, but he has just stumbled into one of the weirdest conversations he may ever know. It is about life and death in the universe, and whether this little child might be part of the first generation to discover it on a planet other than Earth.

I was sitting with Dr Louisa Preston, a freelance astrobiologist, TED Talk supremo and, well, expert in lots of impressive things to do with space. I had been introduced to her at a book launch a few months before and she'd been one of those interesting people I thought I'd quite like to keep in touch with for zoological purposes such as this. 'What do you do?' I had asked her at the party. 'Well, I search for life on other planets,' she had replied. Conversations like this do not happen to me very often – I felt like I had won some sort of competition. 'What planets?' I'd said meekly. 'Mainly Mars,' she had said, casually. She told me that she was most famous for using infrared light to excite leftover organic molecules in rocks, unveiling the tell-tale signs of once-living organisms. She hoped one day to use this technique on the rocks of Mars, to see whether such biomarkers really are or are not present on planets other than ours. She'd be a perfect place to start the book, I'd thought when we met. Perfect in a number of ways. Louisa had spent her career considering all forms of life so she'd be good to talk to about definitions of life and death. Might we expect death on other planets that harbour life? I wondered.

Her perspective on life and death would be so universal (in the truest sense); a world away from all the Victorian body parts in jars that I had been forced to politely inspect whilst drinking my coffee at Death Salon a few weeks earlier. I emailed her and asked if we might meet in the Natural History Museum, London. This was, surely, the perfect place for a journey like this to begin. In its collection are 80 million specimens, and all of them are dead.

We had sat laughing and talking animatedly before Harry, the little boy, had arrived. Louisa had been very keen that before tackling the big question of 'WHAT IS DEATH IN THE UNIVERSE?' we first discussed what, exactly, the definition of *life* in the universe should be. To Louisa, this was (and is) the vital bit. The vital question. It was like a palate cleanser before the main course. In fact, just before we were interrupted by the little boy and his grandfather, Louisa had been trying to question me on whether I believed that a mule was alive or not. 'Of course a mule is alive,' I had said bluntly. 'Of course,' she had agreed, smiling wryly. 'But a mule is missing one of the central definitions of life because it is unable to replicate.' 'Oh,' I murmured. 'People say a definition of life is reproduction,' she continued. 'Well, a mule is sterile. It's not alive.' I gave her a withering and disgruntled 'OH DON'T GIVE ME ALL THAT PHILOSOPHICAL CRAP' face, which she read masterfully. 'C'mon,' she laughed. 'It's true. One of the classic criteria of life is that it can reproduce, regulate itself, metabolise, grow, move, excrete. Mules can do all of the things on this list, but they can't reproduce. So does that make them not alive?'

I gathered myself in the moments Harry, the little boy, was thrust momentarily into our care. He really was going to find this conversation more than a little strange,

I realised. Only a minute after his arrival, Louisa had me justifying a long-held belief of mine that my refrigerator is definitely not alive. 'Why not?' she said, clearly enjoying herself at this point. 'Fridges can regulate their temperature. Thermostats respond to changes in the environment in much the same way that a living thing would. By some definitions there are those that would say that your fridge is alive.' I managed to pull a face that this time said to Louisa 'My fridge is definitely not alive.' Louisa looked puzzled at me, unable to read my face. She continued in this vein for a while, naming other examples of things that some would justify as alive, but that, like fridges, are definitely not alive. 'And fire!' she laughed. 'Fire's a great one!' Louisa particularly liked talking about fire, it seemed. I caught the little boy giving us another fleeting glance. Louisa didn't notice. His grandfather had barely moved in the long queue. 'In every sense fire should be a living organism,' she said. 'It grows, it eats, it reproduces, it spreads, it regulates itself – lots of its products are maintained within the flame, almost like how cells work. So what makes it not alive?' I waited a few moments, silent, waiting for her to continue. I realised then that she was not being rhetorical. 'Why *is* fire not alive?' I heard myself say. There was some silence at this point as Louisa waited for an answer. I drew breath, thinking about it a little more … 'A FIRE IS … A FIRE IS JUST NOT ALIVE!' I offered up grandly. 'The problem is,' she said. 'when you start down the line of asking about what life is, it's easy to get drawn into the very dangerous realm of talking about a *life force*, or a *consciousness*, or a *vital spirit*.' She shook her head. 'That's a mistake we don't want to make.'

Though it was admirably tackled by Schrödinger, he certainly wasn't the first to ask the (im)mortal question 'WHAT IS LIFE?' In fact it is arguably one of the oldest and most philosophically well-trodden questions in science. Among the first to discuss it were materialists such as Empedocles (*c.* 490–430 BC). They argued that life was caused by an exact and appropriate mixture of 'elements' called earth, water, air and fire (the 'roots of all'). Not long after the materialists came Democritus (*c.* 460–370 BC), who was among the first to promote the idea of a soul – a kind of manifestation of fiery atoms that interact in a certain way to produce what our minds define as 'you' and 'I' – our 'spirit'. Such ideas were chewed upon further by the French philosopher René Descartes (1596–1650), who held that animals (including humans) worked more like machines, assemblages of parts that together produced an emergent property: something you or I might describe as a life and a soul. Descartes's ideas and his more rational approach to the definition of life challenged the fuzzy notion of a soul, or, at least, encouraged discussion of what the notion of 'soul' is and how it might interact with the physical body.

In the centuries that followed other definitions were put forward, including the more recent (and even more rational) scientific definitions to which Louisa had alluded and that many of us (including me) remember from biology school: 'MRS GREN'. Many textbooks say that life is defined by Movement, Respiration, Sensitivity, Growth, Reproduction, Excretion and Nutrition (MRS GREN). And under the rules of MRS GREN Louisa is right: mules might not be considered truly alive and fridges and fire might be (and please, please, no one ask about viruses – things, it seems, get very messy indeed when people start asking whether viruses are alive). Overall, when one considers life, one thing is clear. MRS GREN definitions of life are clearly a bit lousy.

'What definition do you work to, then?' I had asked Louisa. Her response was simple. It's the same as NASA's. 'Life is something that undergoes Darwinian evolution,' she said. So that's what she is searching for: evidence of evolution on planets and moons other than our own. 'How exactly does one search for evolution?' I asked, slightly confused. 'I know. I know,' she laughed. 'For my kind of work, it's a pretty ridiculous definition. I mean, we're hardly going to sit there staring at something that *might* be alive, waiting for it to evolve. How can we test that? We can't really. But it's the best and perhaps most all-encompassing definition we have. So that's what we work with.' And so it stuck. In a weird sort of way, though, this definition of life being Darwinian in nature pairs nicely with Schrödinger's when it comes to life and death. Both life and death are inherently linked to one another, after all. Life is a flamboyant and whacky conveyor belt that ultimately moves things endlessly from a state of borrowed order toward a state of chaos. And life breeds more life through Darwinian evolution. 'So if that's your definition of life,' I said to Louisa, 'then what is your definition of death? How do you define death in the universe?' The question hung in the air. There was a brief pause. Harry squirmed. Louisa thought about it some more. 'I don't know,' she admitted. 'I guess I don't really think about it that often.' This surprised me, a little.

So how can we consider death, exactly? What *is* death? How might we define it? By ploughing into this question we find ourselves, frustratingly, back in that same rats' nest of definitions. Most agree that death describes the state that exists when life … runs out. But I find this too woolly a definition. For instance, for the Victorians a stopped heart meant you were dead. It was as simple as that. But now it doesn't. Because of defibrillators, a stopped heart is only a symptom that can be potentially fixed, provided too much time hasn't elapsed. Death hasn't occurred. Another example is with drowning: there was a time when a drowned man or woman was considered definitely and most certainly dead, but then, once CPR was hit upon, they weren't. In both of these cases death was a term used when describing something lacking any remaining potential for life. But this isn't fixed or written in stone.

Nowadays we consider death as being a fixed, easy-to-label state, but actually our clinical definitions aren't that much further on from those of the Victorians. Even today occasional medical cases show that our definitions of death may be wrong and need changing. In the late twentieth century those in a vegetative state were still considered by many to be 'brain dead' – yet those in such a condition can continue to grow and develop, and even give birth. So the term 'brain dead' is clearly not accurate any more, and not appropriate either. Things we consider 'dead' now may not always be so, which is a fascinating thought. But allow me to consider this thought experiment with non-human animals. Consider those creatures that undergo cryptobiosis, able to survive for long periods as lifeless shells or hardy eggs. Consider, for instance, the tiny sexless metazoans that live in birdbaths (among other places), the bdelloid rotifers, which expel all water from their bodies and form a hard stone-like ball when their puddles dry up. Think about them. They can last for seven years in this dehydrated state. They undergo no growth or metabolism, nothing like that, in all that time. They are surely not alive in this state … but they are surely not dead either. They might revive. And then there are the sea monkeys (brine shrimps), which can undergo cryptobiosis like bdelloid rotifers but for far longer, perhaps for centuries in some cases. Not all of these dehydrated life forms will find water. Many of them may blow away or be buried in places without water, and many will break down over years or decades, eroded by the elements. Entering chaos. But at what point do we say that these creatures have died? At what exact moment could we call the time of death for a dehydrated seed such as this, as a doctor might? This is surely impossible. It's hard to escape woolly definitions, it seems.

'HARRY? HARRY!' A loud voice broadcast itself over the busy museum cafe and sliced through our conversation. 'OVER HERE!' The grandfather returned carrying a tray with a cafetière, a soft drink and some biscuits. He'd been ages. I realised at this point that we had both totally forgotten about Harry. I looked immediately to the seat next to us and was thankful to see him still sitting there. He was swinging his legs underneath the chair, looking a bit solemn and pretending not to listen to us again. 'Harry, *this* table,' said his grandfather, gesturing him to leave us and come over to his table nearby. I gave the grandfather a nod and a polite wave. He smiled and gave us a little thumbs-up. The little boy slid himself off the chair and went to join his grandfather. I wondered what in God's name this child was going to tell his grandfather we had been talking about.

Louisa told me more about her work investigating Mars, describing it as a new and exciting frontier in the search for extraterrestrial life. 'Mars is accessible. It's where we can search easily, but also it's had a history like Earth's. It had an atmosphere. It had water. It had an environment probably much like the early Earth.' 'But should we really expect to find life there?' I asked. She shrugged. 'Life arose here on Earth,' she said, 'so there's really no reason why it couldn't have evolved on Mars as well.' Louisa clearly loved Mars. Her voice was full of awe and passion and she started to talk much more quickly whenever it was mentioned, which may have been the coffee, but I really didn't think it was. 'The geological processes during Mars's history are so different to ours,' she said. 'It doesn't have plate tectonics, for a start. It hasn't destroyed its history as happened so often on Earth. We can look at the rocks from the beginning of Mars's history and they could still preserve fossils or evidence of early life on the planet.'

She told me about the ExoMars rover, which goes to Mars in 2018 with the sole mission of finding life. 'It could be that in the next few years we drill into the surface of Mars and find that protein,' she said. 'That crucial evidence: evidence of life outside of Earth.' 'And what do you think everyone will say if you find it?' I asked. I have often fantasised about that moment, as we all have, I think. When we realise, as a species, that we aren't alone in the universe.

'Well ...' she said. 'I think the public will probably be incredibly disappointed.' She sighed. This wasn't what I had expected. I had expected Louisa to have entertained ideas of global celebration, peace and unity spreading across humankind; an acceptance and early understanding that we weren't alone and that we should look after each other a little better. 'Why will everyone be disappointed?' I asked. She shook her head sadly. 'They'll be disappointed because everyone wants to find life on Mars, and then we'll stand up in front of a global audience and say "So everyone, we found a *fatty acid*!" ... and they'll be like ... A *FATTY ACID*? No bugs? No aliens? ... A FATTY ACID?' We both laughed.

It struck me at this point that the popular view of astrobiologists is that they are searchers for distant life, but in reality most of Louisa's job is about looking for death. Evidence of former life, through fossils and fossil biomarkers, rather than life. I mentioned this observation to her. 'That's actually quite interesting,' she agreed. 'Yes ... death is essentially what we're looking for ...' She looked into her empty coffee cup. 'I've never thought of it like that. But you're right, death is almost certainly what we'll find on Mars.' She continued to stare at her coffee cup. 'I guess, yes, all I study is death, really. Fossils ... remnants of dead organisms. All of the building blocks of life I look at are what's left after something has lived, died and degraded – been destroyed, basically.' The disorder again.

We sat in silence for a few moments. I looked over at the little boy. He was talking animatedly to his grandfather, who was listening intently to what he was telling him. Grandfather looked interested but there was a tiny bit of concern and seriousness in his eyes, as though he was deeply worried about what Harry had been describing to him at that moment. It dawned on me that Harry could be telling his grandfather that his fridge was quite probably alive and that the thermostat is unable to produce sterile offspring. But then another thought crossed my mind. I wondered if it was the first time that anyone had discussed death in front of that little child. Hardly anyone speaks about death in front of children. Had that moment with us been Harry's first time? And I wondered if what we'd talked about had damaged him or whether, perhaps, it had done him good. It was a thought I knew I would come back to. And I did a few months later.

Louisa and I said our goodbyes to one another and promised to keep in touch. She wished me well with the book. I wished her well with finding new life on other planets. It had been a useful place to start, on the whole. Hearing about Louisa's work had been fantastic, but finding a clear definition with which I could begin my journey had proven quite hard. As Supreme Court Justice Potter Stewart famously said of pornography, 'I know it when I see it.' … Well, I guess I was left to think this way about death. We know death when we see it, in much the same way we know life when we see it. On the train home, I thought about where to go next. And then I realised something. Louisa had been a little surprised when I had mentioned that, really, she studied death in the solar system and not life. Well, in a funny sort of way, I was the opposite. I had assumed I would be studying inert things whilst researching this book, dead things like the fossils in the museum, but really I would be studying life. The *potential for life*. To understand why frogs, on the whole, live for a few years, I would need to look at living frogs. To understand why spiders occasionally kill one another I would have to look at living spiders. To understand why some jellyfish apparently manage near-immortal feats I would have to understand the life of a living jellyfish. I would have to see living things and speak to living scientists. And suddenly I got it. This wouldn't be a book about death at all, I realised. It would be a book about life. I would be studying life. I would be reporting on life.

This would be a life story. A *proper* life story. And in the bits in which it wasn't a life story, when death was inevitable or had actually happened, the spectre of potential life would still lurk, even if it was in the form of other creatures like worms, blowflies or scavenging foxes.

CHAPTER TWO

Senescence and What Waits for the Lucky Few

'He's going to be a little bit floppy here, just be careful.' The vet deftly hauls the bulky fish from a water-filled tub on the floor and gently places it upon a foam bedding on top of the operating table. It wriggles slightly but seems quite calm. It is about the size of a salmon. According to the aquarium it's a copper rockfish. There is activity all around: technicians gather around the table on which it flaps in its foam bedding. There is some to-ing and fro-ing in the background of the shot. Someone places a hose into the fish's open mouth to keep the gills oxygenated, while someone else has the simple task of keeping the fish's skin nice and moist. This looks an easier job. Yes, the fish seems surprisingly relaxed about all of this attention. Someone clamps it in place on the bedding. It looks less relaxed now; it flips and waggles a bit. Ready to operate, one of the two lead vets pulls the overhanging lamp up close to the fish's cold head and gets ready to perform.

Where its eye should be, the fish has a gaping pink eye-socket, which recedes deep into its skull. It is missing its eyeball. The vets look anxious. Suddenly a host of bystanders appear at one end of the room – this is what they've been waiting a while to see. A vet is going to do something to the fish with one eyeball, but the viewers (including me) aren't yet sure what that might be. They quietly watch the two lead vets do their thing. I am sitting and watching this miles away on my laptop. 'So …' says the male vet, holding what looks like a needle and thread. 'So where do we …?' He gets up close to the eye socket and looks very closely at it. 'Do we do him there?' He points to the top of the arch of the socket above the missing eyeball. 'Or there and there?' He points to another part of the socket. The female vet leans in. 'I normally put it up against the fish first …' She takes an object from her hand and places it into the fish's empty eye-socket. It is a cheap-looking plastic googly eye. She arranges the cheap-looking plastic googly eye neatly into the socket. 'Nice job,' someone says quietly in the background. The operating room seems increasingly tense and there is definitely a hint of excitement in some of the bystanders. The male vet continues: 'So I just use a needle to make a hole through the bone …' He carefully moves the needle, pulling through the thread. 'And then I try to run the suture through the same hole.' Someone in the background looks slightly nauseous at this point. Water continues to pump via a tube through the fish's gills. The fish shudders ever so slightly. Minutes pass. The video cuts forward. We are near the end of the operation now. And then, 'So … we're cutting the suture.' The male vet remains calm under the pressure, narrating each movement. Almost there … 'Getting rid of the little tags …' Almost there. 'And … nice,' he says.

Everyone admires his handiwork. He looks pleased. And he should be. He has just successfully sewn what looks like a plastic eyeball onto the side of a fish's head with the skill and dexterity of a young Jim Henson. And this really did happen. The unusual operation was undertaken by experts at the Vancouver Aquarium, and the video of the operation is freely available online for all to see. It's interesting. Interesting because it tells us a little something about animals and the rarity of their reaching old age. Why did this copper rockfish need a googly eye? The answer is that it had developed cataracts. It was old. The cataract had occurred in only one of its eyes, resulting in that eyeball being surgically removed, but this hadn't worked out well for the rockfish. Other fish in the aquarium had been observed to 'bully' the one-eyed rockfish (which they are rather prone to do), so Vancouver Aquarium decided to do the right thing and reached for the needle and thread, and the Hobbycraft Multipack of Googly Eyes. Cataracts are one of many diseases of old age. And, kept away from predators, aquarium fish age in many of the same ways that we do. Cataracts can be quite common. Ageing is natural.

So why do we all age? Why don't we just carry on living in perpetuity? Why can't we be immortal? It is one of the questions that has plagued humanity for centuries. From ancient Greece's Tithonus – requested to be made immortal by Zeus without anyone throwing in the bit about eternal youth – to almost every Hollywood star you care to name, our obsession with ageing is absolute and (ironically) timeless. So *why* do we age? Why can't cells just continue to replace themselves in a predictable and healthy manner forever? Why do animals fall apart in old age? And why do cataracts often feature as a symptom of old age in animals as seemingly unrelated as fish and humans? Dig about in the science of senescence (as ageing is called) and you'll quickly come to realise that the answer is, almost spectacularly, that no one is entirely sure. Not that this really matters; as with all of zoology's known unknowns, scientists over the last century have gathered into camps supporting a host of different hypotheses that may explain what might be going on, particularly at a cellular and molecular level, to produce the phenomenon we call ageing. There are many hypotheses out there, and I will outline the three most common nice and early in this chapter before returning to them in the course of this book.

The first hypothesis to explain ageing is that cell damage simply builds up over time. As generation after generation of new cells is created within the body, accumulated DNA damage occurs that somehow affects cell renewal. These errors build up. Bit by bit, organs fail to repair properly. Bit by bit, cells become dysfunctional and the diseases of old age result. The second of the main hypotheses for ageing relates to 'free radicals' – the highly reactive atomic particles generated particularly when mitochondria (the biological battery-packs present in each of our cells) manipulate oxygen to fuel the energetic reactions required for life. Free radicals have the potential to stress a cell's function, that much is certain. It may be that they lead to an accumulating cellular burden with age, much like the first suggested cause. And then there's the third hypothesis: that of a telomere. As you will know, all of the cells in your body contain chromosomes in which are housed your genetic blueprints. The structure of these chromosomes is interesting: each of them, when pulled apart, is capped at each end by a 'telomere'. Telomeres are special lines of genetic code which act a little like the plastic protective tips at the end of a shoelace. Experiments suggest that, for each successive division of a cell, the telomeres shorten. This puts a finite limit on cell division. The hypothesis is that, somehow, this shortening puts a kind of cap on the number of replications a cell can undergo, potentially limiting the activities of bone marrow and arterial lining, where repeated division is conducive to sustaining life. The result, as with the other hypotheses, is cellular breakdown: what you and I call ageing. There are certainly other possible explanations for ageing, but these three (accumulating DNA damage, free radicals and telomeres) hog much of the limelight, partly because they have been studied the most intensively.

Research into senescence is anything but an unappreciated zoological backwater. Understanding ageing is becoming a key battleground for those from a host of scientific disciplines, many eager to tackle diseases of old age that include cancer, cardiovascular disease, arthritis, osteoporosis, type 2 diabetes, hypertension, Alzheimer's disease and cataracts. These are natural diseases in many ways. But how could natural selection have produced them? Could there really be a point to these diseases, or are they simply by-products of some other process? This question has rattled scientists for a generation.

Among the first to offer a Darwinian perspective on animal ageing was the British biologist Peter Medawar. Medawar became the so-called 'father of transplantation' through his pioneering work on immune tolerance and organ transplants. He understood that every animal alive at any given time has a particular chance of dying. In fact, this probability of death was integral to his theorising about ageing (senescence) and why animals age like they do. Medawar's 1951 inaugural lecture at University College London, *An Unsolved Problem in Biology*, went on to underpin all three of the modern theories for senescence by flipping the problem of death on its head. Rather than focus too heavily on any given species, he asked why it was that natural selection didn't 'cure' all animals from ageing and dying. Natural selection is adept at solving problems, right? Yet death is the biggest problem of all, surely. So why is death so prevalent in the animal kingdom? he wondered. Why hasn't it been whittled into shape by natural selection, as one might expect?

Medawar was the first to appreciate that the answer lay in probability. He realised that, even without considering ageing, there is a statistical likelihood of death for any creature at any given moment. All animals will die eventually; that's 100 per cent likely. Some animals might die when still relatively small and young, when they become easy prey for something else. Some species might be more likely to die once mature and in competition for mating resources (like nests) or searching for mates. We all have a chance of death, and this probability changes during our lifetime. In humans, statistically, once we reach 30 years of age our chance of dying doubles approximately once every eight years. It's as simple as that. It may seem obvious to us now, but in 1951 this idea of probability was an important insight. Medawar understood that for every second that a bacterium swims in a pool of water, for instance, there is a probability of death that glows like a neon sign above its head. The given bacterium may die through predation, cosmic radiation or misadventure, and plenty else. It could be besieged by virus particles. Squashed or dehydrated into oblivion. Every day that a bacterium lives it is rolling the dice, unaware that one day its numbers will come in. And they *will* come in eventually, no matter how expertly that bacterium piddles about in that particular puddle of water. Natural selection, Medawar realised, simply favours the animals that get on with the business of reproduction before their chances of dying increase. An organism in that puddle that breeds every two days, for instance, will far outcompete an organism that breeds every two weeks or two years. Simply, gene pools become flush with those best primed for reproducing before the statistical chances of death increase.

At its most simple, Medawar's idea was that natural selection didn't sort out senescence because it was drawn to the battleground of life; it was drawn to sex. The only language of transmission it knows is through reproduction. And those sexy genes, high on life and sex, think nothing about solving the problems of old age. They think only of carrying on. As ageing individuals we, according to Medawar, have been left high and dry by our genes' insatiable desire to spread at all costs before death gets us. In nature, when death doesn't get animals early, the lucky few that remain will see senescence expose itself to pick them off anyway. And that googly-eyed rockfish, kept safe in its tank, is just that. Like us, it lives a different kind of life. One, on the whole, without predators. Medawar's idea was a catchy one. It influenced a generation of scientists with it, but doubts still remain about whether, and to what degree, his idea works in the real world. Could senescence really be as simple as Medawar described? Could the signs of ageing really just be the result of sloppy maintenance that can't be fixed by an unthinking selection machine obsessed with sex and cutting-edge replication? We still can't be totally sure.

In 1966, the evolutionary biologist George C. Williams offered up another way in which senescence may appear naturally in a way that is similarly unfixable by natural selection. He proposed something called 'pleiotropy' – a situation where the flipping of genes on and off can cause multiple, seemingly unrelated traits to expose themselves, either instantly or later in life. To Williams, the problems of old age come about as side-effects of things that help animals have better sex earlier in life. Williams's example was to imagine a hypothetical gene that alters calcium metabolism in a way that both strengthens bones in youth and occludes arteries in old age. In this situation, natural selection would be drawn to maintaining the healthy strong bones of fertile youth – those genes would spread, regardless of the consequences later in life. In many ways Williams's view works in the same way as Medawar's. Sexy genes spread better than genes for healthier arteries in old age, which barely spread at all because of the fertility drop that occurs with ageing. Natural selection can barely touch the diseases of old age, let alone try to fix them, and so these diseases remain. 'Reproduction is the beginning of death,' wrote James Joyce. And so it is. Sex and death really are two sides of the same coin.

And so the googly-eyed rockfish lives out its life. Its wonky eye is an expression of a fleet of genes obsessed with sex, which barely gave a nod to trying to fix a common disease of old age. Stare closely enough and you will see yourself reflected in that googly eye. The universe has given you a statistical chance of death, and natural selection has acted accordingly on your growth rate, your brain development and your puberty. Each is programmed with sex in mind, not cosy senescence.

Does this make you feel small and helpless and a bit insignificant? I confess that exploring all of this does make me feel a little helpless about ageing. But don't feel too glum, because there is a silver lining and it comes down to a simple observation in nature. It may not *always* be this way for us. Many populations of a given species appear to evolve longer lifespans quite quickly in evolutionary terms, particularly in habitats that lack predators. Give an opossum an island habitat with fewer predators, and a couple of thousand years later the opossums will live twice as long, ageing at half the rate of their mainland competitors. Their lifespans change. Brook trout offer an even more spectacular example: their introduction into cold, nutrient-poor waters in California's Sierra Nevada has seen their populations quadruple in lifespan. Where once they lived six years, now they can expect to live for 24, with the only apparent catch being a delay in sexual maturation. Fiddle with the statistical chances of death and animals evolve, not within millions of generations, but hundreds. The writing is on the wall. Senescence isn't a phenomenon etched in stone. Senescence is fluid. Natural selection, if needed, can pull the strings of ageing, which means that there may be a genetic component to ageing. The brook trout and the opossums offer us a startling revelation: diseases of old age – *all of them* – might be something we can delay. And by fiddling with genes, we may discover that they can be delayed greatly. Perhaps immeasurably.

The study of ageing really is far from a scientific backwater; it is recruiting geneticists, zoologists, biochemists, molecular scientists and physicists. They are queueing up, eager to weigh in with their own given insights. And why? Because they are drawn, like so many humans before them, to the prospect of immortality. But this time it's real.

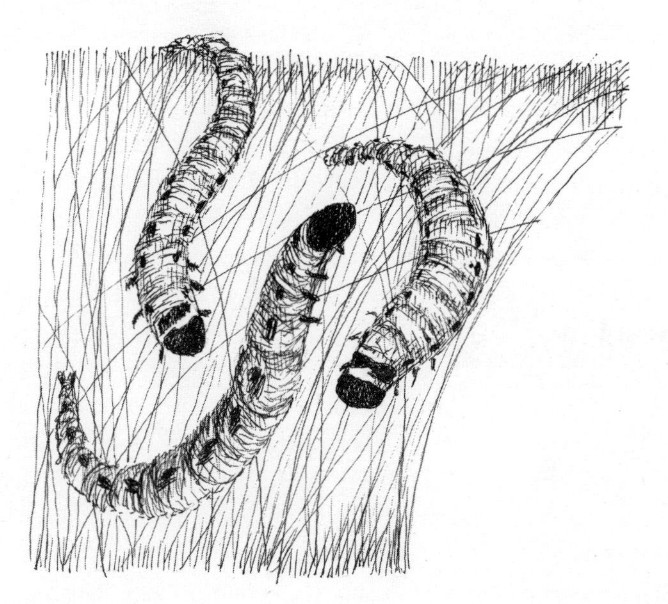

CHAPTER THREE

Fear and Loathing in Birchwood

'Anyone want a dead magpie?' her message on Twitter had read. My eyes lit up. Me, I thought. I want a dead magpie. That's *exactly* what I want. I really did. Genuinely, I really did want a dead bird. I had quite wanted a dead magpie or crow as part of some initial research I was undertaking on how corvids (the family which includes crows, jackdaws, magpies and jays) respond when finding dead members of their own species. My plan was simple: I would get my hands on a dead magpie or crow or jackdaw and rest this corpse gently in the middle of my local corvid haunt, a spinney in a nearby field in which a jackdaw colony rubs up against small flocks of crows and magpies. What would they do when

I placed the magpie on the floor? How would they behave? Would they inspect it? Become anxious? I was interested. Sure, it wasn't exactly a randomised controlled trial at this point but it was a start, I thought. I had to have this dead magpie. I *had* to have it. I'd *travel* for this dead magpie. So that is exactly what I did. I travelled to Birchwood to meet Alison Atkin, bone expert, archaeologist and self-anointed 'deathsplainer'.

Birchwood is near Warrington, which is midway between Liverpool and Manchester. I had never travelled to Birchwood before. I like going to new places – it was quite exciting. As the train pulled in it looked a clean, ordered place. Birchwood is what we in Britain call a 'new town' – a place without historical infrastructure, where county planners in the 1970s could work unimpeded by irregular angles laid down by the Victorians. So Birchwood is a place where everything sits at 90 degrees, and into which paving slabs are perfectly flush up against each other, which makes the place look very neat and tidy and organised. There are cycleways and pedestrian zones in Birchwood. Lots of seventies-style flats and other rectangle-shaped objects. And tree-lined avenues, which is something you don't often get in Britain, given that trees take up space and space is at a premium in many parts of this relatively tiny island.

I walked along the station not knowing quite what to expect from Alison. Would she be like the people at Death Salon? Would there be the same air of mortician chic in her? Would she approve of what I planned to do with her dead magpie? I really didn't know because our correspondence hadn't been particularly detailed up to this point. In fact it wasn't detailed at all. It was: 'Who wants a dead bird?' [PAUSE] 'Me.' And that had been pretty much it. We'd arranged it all, in fact, via Twitter. I had wondered whether Alison did this often, clandestinely meeting strangers at train stations and handing them dead things. My palms were sweaty. I looked up. There was only one person waiting in the arrivals section behind the turnstiles and it was, I guessed, Alison. Thankfully, she was smiling and she looked friendly and normal. She had a bag in her hand – it was an enormous plastic carrier bag in which I imagined the dead bird lay. On the side of the bag were capital letters encouraging supermarket shoppers that the bag be reused. The letters read 'TESCO: BAG FOR LIFE'. I wondered if Alison – a Canadian – was demonstrating to me that she possessed a true sense of British irony. She had a wry smile on her face. I think she did.

It had proved quite hard to predict how long it might take to hand over a dead magpie. What's the etiquette? I'd wondered. In my ticketing plans I'd allowed myself two hours between getting the train to Birchwood, taking the magpie off Alison and leaving Birchwood again, but actually, as it turned out, the hand-over had taken a little over 60 seconds. Having about 119 minutes to kill, we decided to walk the streets of Birchwood and talk more about her work and the world of zoological death that I was now entering. I felt that Alison was going to prove useful to me – I was after all a relative beginner in the field of dying – but I wasn't yet sure how. I wanted to know about Alison's experiences having an academic career that involved, in every sense, death. I wanted to have what I knew she had: a rational approach to discussing death that is underpinned by science, but that isn't graphic, crude or cruel in its delivery.

On our way to the coffee shop I told Alison about my publisher's understandable anxiety about me writing a book about death, and how there was concern that people might not like to read about death and dying, or buy books about it. She chewed on this for a few moments in silence before looking at me: 'Death happens,' she said bluntly. 'That's pretty much what I spend most of my time reminding people. Death happens.' 'Right,' I said, jotting this down. Death happens. Alison is an osteoarchaeological researcher, which means she is interested mostly in digging up human remains (bones) and studying them. She tells me that her main research interests include mass fatality incidents such as the plague, and how she was specialising in looking at methods to identify the hundreds or thousands of unidentified or misidentified bodies buried after such catastrophic events. The burial sites after mass fatalities are often chaotic in their structure, she told me (something that is still the case today in cases of genocides or sudden disease outbreaks), so Alison is trying to work on methodologies to unpick the chaos, as it were.

We sat down for a coffee in a busy coffee shop not far from the station. 'You ever work with kids?' she asked me as we sat down. I nodded and considered mentioning Harry, the little boy from the museum. 'A bit,' I replied. 'Kids love death,' she said thoughtfully. 'They love it. You bring skeletons into schools or museums and kids *surround* you. They have so many questions. They want to know "*Where do you get bones?*" or "*Who was this person?*" and you can see their parents wincing a bit in the background.' That's what I felt the grandfather had been doing. Sort of … wincing … when I saw Harry detailing to his grandfather the conversation he had overheard from us about death. 'And do you tell the kids who the bones were from?' I asked. 'Yeah,' she said. 'I don't dress it up. I don't use terms like "passed on" or "passed away" – nothing like that.' Her open face had turned suddenly more focused and serious talking about this. 'When I work with children I use "death" and "died" and "dead" and "dying",' she said. Each word was clearly spoken. 'Label it properly,' she said, offering me advice. 'Frame it properly, otherwise you risk miscommunicating death. And that can be a big problem.'

To Alison, there is a big problem with society, and that is that we fail to talk openly about death. This, she told me, was where the whole *deathsplainer* thing came in. Alison was trying to alter the zeitgeist a little by encouraging frank conversations about the bit that happens after we live. 'It's important we talk more as a society about death,' she explained. 'When a kid loses a pet do the parents allow their kids to hold it? Or even see the dead body?' I thought about this. 'No, I'm not sure they do,' I answered. 'When a kid sees a squashed frog,' she continued, 'they want to ask about it and we don't always want to talk about it but … I dunno …' She paused. She held out her hand as if she had an imaginary dead frog in her palm. 'LOOK!' she said. I looked at the imaginary dead frog in her palm. 'THE FROG IS DEAD!' she said assertively. I nodded my head. 'That's what we should be saying to young people: THE FROG IS DEAD,' she repeated. I thought again about Harry. And I thought about my own kids, and how we had avoided talking about death to them, sometimes choosing instead the awful '*It's only sleeping!*' line when coming across a dead creature, something I realise I am deeply embarrassed of when I come clean to Alison about it. But it had made me wonder, as rather a bastion of science and rational thinking, why on earth I'd said this. Why hadn't I been clearer from the start with my young kids? It wasn't sleeping. It was dead. Alison was quite forgiving about all of this. 'So many people say that to young kids,' she said. Perhaps people like me avoided talking about death with their young children because we ourselves were scared shitless of it. Maybe it really was as simple as that. We're scared and we don't want them to be scared.

The conversation moved on, and the subject changed to dead birds. 'What sort of person gives dead birds away through Twitter?' I asked her. The nonchalant way in which she had offered up the magpie online had made Twitter seem like some sort of zoological version of eBay or Loot – as though she offers up dead animals all the time. 'Do you?' I asked. 'Do you offer up dead things all the time?' 'You wouldn't *believe* the appetite for dead specimens out there in academia, and the importance of dead things to reference collections,' she told me. 'We have freezers in the basement of our department full of dead fish, full of small mammals, dead birds …' She paused and looked up at the ceiling. 'I think there was a dead badger in there at some point.' She mumbled this last bit very quietly. I tried to picture her freezer. I'm sure each specimen is well labelled and it's delightfully clean and well ordered, but in my mind's eye all I can picture is a gruesome chest freezer from some gory eighties movie. I later ask my scientist friends about the weird dead creatures they, like Alison, have temporarily stored in their own freezers. Their responses include wrens, hedgehogs, shrews, mice, squirrels, field voles, bullfinches, rats, snakes, ants, polychaete worms, leeches, gerbils, an echidna and a muskrat, as well as larger items that would horrify even the most academically weathered spouse of a zoologist. I can only imagine the face my wife would pull if she opened our freezer and saw the head of a coyote, a porpoise torso or a neat row of brains from several species of monkey, for instance (for those interested, the best I've managed in my freezer is a long-eared bat).

We finished our coffees. 'Wanna see something weird?' she said. 'Always,' I responded, without skipping a beat. What the hell, I still had an hour to kill. But what Alison showed me a few minutes later was no time-filler. Alison swerved me into a new zoological realm, one I would otherwise never have thought to investigate. And through it I saw a different side to death. It was an experience that carried me somewhere else. Somewhere I'd never been. Something I'd never seen. And it spun me off, now really and truly, on the beginning of my journey …

'Wanna see something weird?' Alison had asked. Five minutes later we were around the corner from the coffee shop looking at a silken tunnel that had opened up before us like some sort of nightmarish cave. We were indeed looking at something weird. This used to be a tree-lined avenue, Alison had informed me. Now it was something else entirely. The whole avenue had been totally and utterly wrapped in silk. It had become a white tunnel; it was like a winter wonderland a few months too late. I was gobsmacked. 'This all just turned up a couple of weeks ago,' Alison said, enjoying my comic-book response to this absurd spectacle. 'Isn't it just the most wonderful thing?' she laughed. I muttered an expletive at this point – that was the best I could manage.

We wandered into the tunnel. I was speechless. You could almost park a bus under the silken cloak suspended from the tops of branches overhanging the pavement. Great white curtains of the stuff; sheets of silk. But this was no PR stunt or artificial outdoor art installation – this was a stunt created by nature. It was a stunning thing to see up close. What was this natural spectacle doing in Birchwood, of all places? We were walled in by it. It was everywhere. And caterpillars, not spiders, had apparently produced this spectacle. Carefully pulling the silken web apart I could see them – there were hundreds of thousands of black and green caterpillars, each tiny, maybe a centimetre or two long. There were so many that, in parts, their combined weight made them sink through the silk in great netted blobs that dangled from the ceiling like stalactites. Hanging like this they looked unnervingly like socks full of maggots.

We walked among them, these larval dollops, my mouth hanging open in disbelief. It was mesmerising. Certainly it was one of the strangest things I think I had ever seen in my life. And it was in Birchwood, England. It really was like a novelty winter wonderland. I pulled and poked at the silk again. The threads had become so intermeshed that they had formed a solid latex-like sheet made up of hundreds of thousands of individual drag lines left by the solo explorations of thousands of individual caterpillars. So many journeys had been made by so many caterpillars they'd created this sheet, a mesh – now it was a kind of fabric; a solid sheet so condensed that it felt and looked a little like white rubber. This backstreet in the middle of urban Birchwood had been made organic by these caterpillars. Even many of the paving slabs below had been covered in the stuff. Alison and I wandered around, taking photos, awestruck. 'Isn't it just the most amazing thing?' she said again. 'It's amazing that it could just pop up here like this, in the middle of such an urban place. Amazing.' 'Amazing,' I uttered. I still wasn't capable of much speech at this point. Underneath the great sheets one could see that the caterpillars were going happily about their business, stripping foliage from the trees like columns of Roman troops on patrol. Actually, there weren't many leaves left.

Local residents, upon seeing us show an interest in the caterpillars, had come from their houses to join us, asking us if we knew what was behind the curtain of silk, discussing and speculating about this strange army. They were bewitched, mostly. And also disgusted. They were disgusted and bewitched. Their suspicion was that these caterpillars didn't *belong* here. Too *exotic*, they thought. This is Britain, for goodness' sake. This is *Birchwood*. Weird stuff doesn't happen in Birchwood, I could tell that they thought. In some of these people's eyes the caterpillars couldn't be trusted. What would happen when they ran out of food? Would they come for the gardens? they wondered. Tongues were set wagging … For many minutes we poked and probed the strange structure, taking notes and photos. But then, sadly, it was time for me to go. Alison and I didn't say much on the walk back to the station. I think we were both a bit dazed from it all. She handed me the bag of bird.

We said our goodbyes and promised that we'd keep in touch. And we did.

Upon leaving Birchwood I did some research into what we'd observed. What we had seen were bird cherry ermine caterpillars, apparently. Ermines are neat little moths, a wintry white colour with little rows of black dots upon the wings. They are renowned (though I confess I'd never seen it before) for wielding silk like some sort of protective curtain, though protection from *what* exactly is as yet unclear. Silk, evolution's most brilliant thread, was here being wielded in a suspiciously altruistic 'panic-room' form. Safe under the 'tent' from intruders in avenues like these the ermines go about their work, specialists in absorbing the tree's atoms and reassimilating them into caterpillar form and then into adult moth forms that emerge in midsummer. According to entomological accounts, each year various parts of northern Europe become festooned in this silk should the caterpillars' foodstuff (bird cherry trees) have a good year. In an archived Dutch newspaper I found reports showing rows of cars and bikes parked under cherry trees, that had become consumed in the mysterious stuff during late spring. I read elsewhere that one of these vast latex-like sheets appeared and spread itself all over the hawthorn bushes at the entrance to Belmarsh Prison in London, to the confusion of inmates and everyone visiting. Such infestations only occur perhaps once every three or four years and culminate, as they run out of leaves on which to feed before pupating, in the caterpillars spreading far and wide in search of other forms of nutrition.

Sometimes, it seems, the ermines do go on the rampage. 'In Finland I have seen a trail of thousands of larvae marching along a railway line,' says Professor Simon Leather, an entomologist at Harper Adams University on his popular blog *Don't Forget the Roundabouts* ('They didn't survive the passing of the 08:50 from Helsinki,' he reports dryly beneath). Leather also recounts hiding from the rain under silken sheets produced by bird cherry ermine moth caterpillars. What we had seen had been like that. The silk could have sheltered us from the heaviest of storms in that Birchwood street. Hours of rain and I think we would still have been bone dry underneath it.

Apparently the bird cherry trees themselves rarely die from these infestations. It's likely that they can weather heavy infestations like those that we saw, provided they are sporadic, which they often seem to be. On the whole, leaves come back the following year and life continues for the bird cherries. This was good news, and it filled me with a bit of hope. One of the passers-by Alison and I had spoken with had displayed something approaching anger about the whole thing – that the caterpillars had stripped the trees of their vigour, and ruined the look of the avenue in spring. I had heard another passer-by later on calling them 'pests'. Both of these comments had rankled me a little. I'm not the sort to really show it, but 'pest' isn't a word I like to hear used about animals. Plus I understand enough about parasites* to know that the true killers among them rarely last long; the most evolutionarily long-lived parasites

*Am I really arguing that caterpillars aren't herbivorous grazers, but are actually parasites instead? The idea sounds, to a zoological pedant perhaps, like heresy. But, loosely-termed, feeding relationships between animals do occasionally cross over like this. Parasites are organisms defined by having a non-mutual symbiotic relationship where one organism gains at the expense of another organism. Isn't that the relationship between trees and their caterpillars? Caterpillars share many features with classic parasites like many lice and mites. They possess specialised adaptations for feeding and holding onto organisms many times their size. They reproduce more quickly than their hosts. Often they depend on them for survival. In fact many caterpillars have evolved to depend solely on one given food plant for the progression of their life-cycle. Caterpillars tick many of the boxes to be considered parasites. And so I think it's acceptable, in these pages at least, to think of the relationship as parasitic. So there.

are those that take a little, but not too much. So, to me, the ermines weren't pests. No way. The ermines were farmers, really, and the leaves were their crop. I felt almost sorry for them. Yes, you know what? I did. I felt sorry for those little caterpillars. When I remembered standing underneath that enormous snow-white canopy, I could see only life. Yet many of the passers-by didn't see that at all. They saw only the associations with death and pestilence.

And so it happened – two weeks later my fears were realised. I should have expected it, to be honest. I received word from Alison that they were gone – the whole avenue of cherry trees had been cut down. Local residents had complained about the ermine caterpillars – the caterpillars had been deemed a health risk, apparently. The whole avenue of trees along with all of the caterpillars had been chipped. Lost. Gone. Dead. The local housing association gave a statement to the *Warrington Guardian*, and here is what they said: 'The infestation was reported to Your Housing Group by a number of residents, and given the trees' proximity to surrounding residential properties, and the extent of the infestation, the decision was made that it would be better to remove the trees completely.' I had come to Birchwood to pick up a dead bird, but I had wound up embroiled in something else that was to become a deeper part of my story of death. I had rubbed up against my foil for the first time: humans. Humans and our strange and sometimes perversely arrogant attitude to other life forms on Earth. Humans and our total terror about anything that reminds us of our own mortality. It proved not to be my last run-in with this peculiar species.

As Carl Zimmer observes in *Parasite Rex*, there was a time when society viewed parasites as nothing more than troublesome hangers-on. Annoyances. Disgusting evolutionary by-products. But then our attitude changed. With the advent of the Cold War, the *Alien* franchise (among a host of sci-fi parasite films) and (most importantly) a new evolutionary understanding of how parasites function, our societal view began to change. In more recent years, it seems to me, there has developed a grudging respect for the wily perniciousness of parasites, as there should be. But it is respect nonetheless, and I am thankful for it. For there is a new and vibrant ecological understanding of the breathtaking constellations that form when parasites are added into our standard model of food chains and food webs. We overlook their role in life and death at our peril. Quite literally.

Though we consider parasites as agents of death, they are far from it; they are much more valuable than that. They can be harbingers of new life, and that's just for starters. In terms of speciation, particularly, they are likely to blossom upon every new form of complex life that natural selection produces. If an individual species of fish should find itself in a newly separated lake, for instance, each of its many parasites may become wedded to the form into which that fish will evolve. Parasites get swept up in the speciation melee. In events like these, one new fish might result in something like 10 or 20 new parasites, or so it's imagined. Indeed, this speciation parallel between parasites and hosts explains some of the patterns we see in taxonomy. Take this example: add up all of the known mammal, reptile and bird species and you'll come to a figure of almost 25,000 species. Not bad, you might say. But then look at parasitic wasps. There are *100,000* species of parasitic wasp that taxonomists have so far named. One hundred thousand. Ermine caterpillars fall somewhere near this pack, capable of occasional ransacking of specific species of trees to which they have adapted, once every few years. Some individuals undoubtedly try it more but their hosts die, so … well, they die too. The world has filled up, in a strange sort of way, with 'nice' parasites, on the whole. Parasites that seem almost to value life over death. The ermines, then, are no different. Yet we offer creatures like this barely a nod of appreciation or any form of inter-species kindness at all.

You might hate them for what they did to the trees, but don't. The trees probably weren't sitting and taking it. The war between caterpillars and trees is at least 50 million years old, and that particular housing association had only recently blundered into it. Trees have had plenty of time to evolve weaponry to combat caterpillars, plenty of time to fight an embittered cold war. So, please, don't feel sorry for the trees. Many can handle themselves – most have evolved some level of toxicity, for starters. Many plants, including trees, extract minerals from the soil and turn them into allelochemicals, some of which will target independent caterpillars' tastes, killing them or (at least) discouraging them from feeding. Of course, in most cases, a genuine arms war is what has resulted. The caterpillars have fought back or, at the very least, many have wised up. Some caterpillars, for instance, bite leaf veins and allow the toxins to bleed out before beginning to feed. In fact, toxins are why so many caterpillars feed on the edges of leaves: here, away from the larger veins, toxicity is at its mildest.

Of course, trees employ other tricks to push back the invading caterpillar army, some of which are staggering in their unthinking ingenuity. How they do so is largely unknown, but many trees almost *sense* that they are being bitten and they muster up a different kind of chemical cocktail in retaliation to caterpillars – a molecule aimed not at the caterpillar, but at the caterpillar's own parasites; a chemical message that diffuses out of the tree and floats around in the air, acting like a biological flag that waves to passing parasitic wasps, which promptly fly down and lay eggs upon and within the specific species of caterpillars to which their tastes have evolved. What's most incredible is that many trees appear to *know* the exact species of caterpillar feeding upon their leaves, and which exact parasitic wasp they need to draw over to dispatch it. Perhaps this is why ermine moth caterpillars employ such webbing? Not as protection from birds, as some think, but as protection from invading wasps. Perhaps. It doesn't seem like anyone's too sure about this yet, but it really could be that simple (if natural selection works the way we think it does, then predict a family of parasitic wasps to evolve with web-scissoring jaws at least by the end of this particular geological period).

There is more in the tree's arsenal than this, though. Some trees can influence other naturally occurring parasites to take care of their caterpillars for them, and what they do with these parasites is even more impressive than the whole parasitic wasp 'come-get-them' thing. It really is fantastic. Some trees waft out secondary compounds that influence the infection rates of caterpillars with baculovirus – a curious package of viral DNA trapped within a double membrane, which acts a little like a smart bomb in its caterpillar-killing efficiency. It starts with one accidental ingestion by a hungry caterpillar. The baculovirus enters the caterpillar's gut and multiplies within its local cells before spreading via the caterpillar's tracheal and circulatory systems around the body. Wherever the virus goes, more cells are infected. Before long the caterpillar becomes a chugging and industrious virus factory, and eventually it becomes swollen with billions of these virus particles, ready to burst. But it doesn't burst. What the baculovirus does next is chilling – it manipulates the caterpillar's physiology in two ways. First, it stops the development of the caterpillar, inhibiting moulting (one assumes this keeps the skin tight and fit to burst). But the second thing it does is mess with the caterpillar's mind. By fiddling about with proteins in its brain, the baculovirus makes the caterpillar seek out light. Zombie-like, the caterpillar struggles up the tree. Obese with multiplying virus particles, it heads toward the sun. Nearer and nearer the light it heads – upwards, upwards, upwards. Then the pressure within the caterpillar builds … Eventually it's too much … The caterpillar pauses … And then it happens. Its tightened skin ruptures and an enormous pressure is released; a fountain of billions of virus progeny rains down all over the leaves below, covering the leaves and petals upon which other hungry caterpillars feast. The infection starts over, 100-fold this time. The message is simple: don't mess with trees. Trees are like NATO – they don't start wars, but they know how to play the game, and they are probably a bit shady (pun intended) and untrustworthy at times. They are anything but victims, in other words. Yet humans are so quick to consider trees as such.

Parasites, it seems, have a truly unfair reputation as agents of death. Those ermines didn't have to die. And neither did those trees. Together the caterpillars and the trees had created something wonderful. In all of the neighbourhoods in all of Europe, the bird cherry ermines had erupted, including there in that street in Birchwood, and I was trapped momentarily within their silken spell. I was not to know at the time that this chance event would wind me deeper into death's spell, but it did. And I'm still not sure whether it was for better or worse.

It was only when I was sat on the train on the way home from Birchwood that I suddenly remembered the Bag for Life, which had been perched precariously on my lap. I clutched it close to my chest, anxiously aware that I was responsible for getting a dead magpie back across to the other side of England without accidentally spilling it across the floor, onto someone's lap or into an open pram. In every train station we passed I felt the hot eyes of security guards across my person. I breathed deeply and slowly. I hoped I wouldn't have to explain what was in the bag. I hoped I wouldn't have to have it confiscated. I had big plans for this dead bird. It had to get home intact. And thankfully it did.

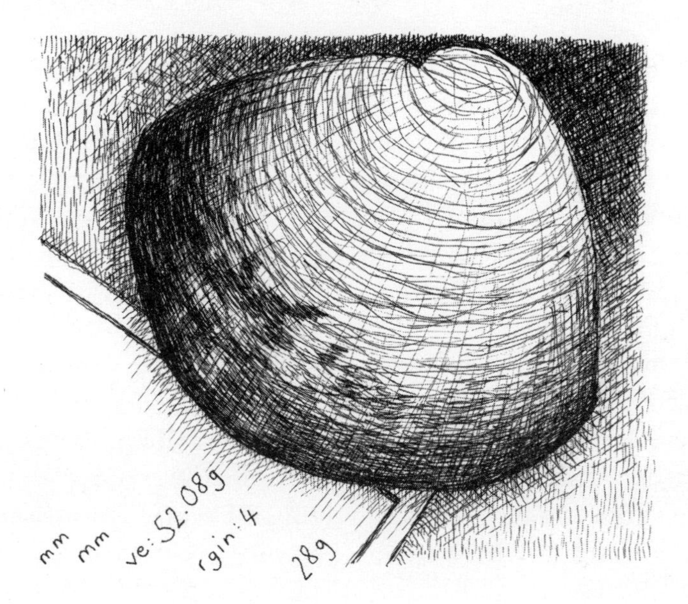

CHAPTER FOUR

Free Radicals and the Secrets Within

Could the rate at which we age really lie in our genes? I was still thinking about the opossums and the brook trout. Might we one day really be able to tamper with ageing artificially? The thought had occupied my mind for months. And then, suddenly, I came across a truly bizarre story – a throwaway anecdote, really, involving humans. The anecdote came from Nick Lane's *Life Ascending*, and it refers to the second of the theories outlined earlier in this book for why we age: the free radical theory of ageing, where highly reactive and short-lived atoms wreak cumulative havoc within cells.

The story is as follows. In Japan there exists a small subset of the population with a common variant in their mitochondrial DNA that alters a single DNA letter, resulting in a tiny reduction in free radical leakage. The result of this variant in mitochondrial DNA is nothing short of spectacular. By the age of 80, recipients of this 'faulty' gene were found to be half as likely to visit hospital for any reason at all, and they were twice as likely to live to 100 years. One faulty gene and they were *half* as likely to visit hospital. One faulty gene and they were *twice* as likely to reach 100. I found this statistic mesmerising. Utterly staggering. Could taming free radicals really be the key to solving the world's numerous healthcare crises? Maybe. There are certainly many animals that do just this. And they do quite well from it in terms of lifespan. In fact, one species does very well indeed …

Like a shrine, it stood in front of us. Preserved in its Perspex display case, it sat atop a shiny black pillar on the laboratory workshop, just like a priceless Ming vase. And this is almost exactly what it is – without price. An artefact from the past. Aged: as old as a Ming vase, in fact. The students and postdocs working the various microscopes had paused when Ming had come out of its cupboard. I got the feeling that Ming isn't allowed out very often. They had paused, almost as if to pay their respects. 'Can I take a picture?' I asked Dr Paul Butler, my guide for the morning. 'Of course,' said Paul. 'Let me put it over here where the light is better.' Paul picked Ming up and took it to the edge of the laboratory next to a sunlit window. He stood back to let me get a good shot. Ming looked resplendent, like an antique polished soapdish, and we both stood in silence taking it in. The uniform striations along its shell looked almost like they had been put there by a comb dipped in paint. Though the millimetre-wide grooves ran in parallel lines across its shell, there were some slightly wider, bleached grooves that looked a little like breaks between tracks on a vinyl record. And to trained scientists, they were being read as such.

Ming was the oldest known non-colonial animal in the world, explained Paul. It had become a global celebrity because of this fact. Ming seemed remarkably indifferent to its relatively newfound status, for: a) Ming is dead; and b) Ming is a clam, *Arctica islandica* (an ocean quahog), lacking in advanced cognitive capacity. Ming was certainly old, though. By reanalysing the growth rings on Ming's shell in 2013, Paul's team determined that Ming had been a staggering 507 years old when it died (previous analysis had estimated an age of 405, which was still record-breaking, but still … 507!).

Ming is now kept in a special cupboard in Paul's lab at Bangor University's School of Ocean Sciences. Most academic institutions become quiet places in the July weeks. Not so in Paul's lab. Throughout my visit, each corridor was filled with the piercing sound of postgrads using lathes to grind mollusc shells into workable fragments for further analysis. Whole banks of microscopes were being manned by students and technicians, each counting and measuring the striations upon their molluscan study specimens. I was there for only a few hours but I could have stayed for days. I really could have. I liked Paul. Having been a software consultant for more than 25 years, he moved from London to north Wales 10 years ago to start a new life as a scientist. He achieved that goal with gusto. Now he leads a team that is using mollusc shells to create a 1,000-year record of seawater temperatures, contributing to our understanding of climate change and how it is acting (and has acted) in marine environments in recent times.

We stopped and chatted to his students, many of whom had their eyes glued to screens midway through their striation measurements. The microscope images of shells were nothing short of beautiful. In some ways, they looked a little like the surface of Jupiter, with a smidgen more order; parallel stripes of cloudy greys, oranges and browns stood out like rainbows on the screen. As in trees, each stripe of colour on these shells represented a single year's worth of growth: they were growth rings, and the size of each is directly influenced by available food, which is itself influenced by sea temperature (Paul's team analyse the shell's chemistry to measure this). By analysing huge numbers of shells, Paul and his students have managed to create a kind of shell-growth archive that provides a way to measure polar sea-temperature change over the last 1,000 years. The ocean quahog shells have become a bit like a climate-change monitoring station; they provide further evidence that the world's climate is indeed changing.

When Ming's record-breaking age was announced to the world in October 2013, newspapers took great pains to put its mighty age into historical perspective. The *Daily Mail* wrote that Ming 'lay on the ocean floor throughout historical milestones such as the English Civil War, the Enlightenment, the Industrial Revolution and two world wars'. Some went with the fact that Ming was born seven years after Columbus discovered America. Others wondered whether its name, Ming, was something to do with the Lib Dem politician Sir Menzies 'Ming' Campbell, who often finds himself chastised for being a shining pillar of longevity, to put it nicely.

For me, well ... I felt none of these things when I spent time with Ming. For me, it seemed unfair to put human history up against it. For there are better ways to add perspective on Ming's life – more animalistic, zoological, interpretations. I found myself expressing awe that, since its 'birth', 16 or 17 generations of my direct ancestry had reproduced and died. Since Ming seeded itself onto that hunk of rock on the seafloor, 169 generations of frogs had spawned and metamorphosed and died, and 507 generations of mayflies had made their maiden flights and then perished. Viewed like this, as a continuum of animal life, Ming is an outlier. But we're all on that spectrum, with our genetic lifespans plotted alongside others of our own species – all of us whittled by natural selection toward a given life history, based in large part on the unseen probability of us living or dying on a given day. Clams fit rather nicely into Medawar's hypothesis that beyond a specific age the evolutionary benefit of longer life becomes negligible. After all, having a thick shell and doing very little in the way of pootling around on the seafloor means that the chances of a clam dying in a single day is rather small (at least until the advent of dredging). Without much death, there's less hurry; lifespan has expanded. But how does a clam take it to such extremes as this? How does a clam get around the problems thought to be associated with ageing? Why don't they expire, smashed and ravaged by decades of free radical abuse, for instance? The answer in their case is that somehow they have managed to tame them. Somehow they have become masters of their free radicals. And they aren't the only ones: many long-lived animals, often on distant branches of life's tree, have evolved the same trick. A trick that one day we may uncover too. A trick that might help us live a great deal longer.

Research into the damage caused in cells by free radical build-up began with an American named Denham Harman. Few people can boast a patent to their name. Fewer still can boast 35 patents to their name. But then Denham Harman came into Shell Oil as a research chemist at exactly the time that the industry in petroleum products began booming. In the early part of his career Harman spent months and years looking for uses for petroleum by-products – essentially he was looking for ways to make money from the waste (interestingly, one of his patents was for the compound used in those yellow plastic strips designed to kill flies). Of particular interest to Harman was what happened to the free radicals produced in and after reactions with petroleum products, and it was this interest that pulled him away from industry and into the arms of academia. Harman was drawn toward the idea that the free radicals that leak from energetic reactions within mitochondria somehow build up, unleashing chaos much further down the line. He had little to go on at this point; it was a hunch, really. These free radicals – essentially rogue atoms missing an electron – roamed cells, in his view, pairing up awkwardly with other atoms. Sometimes these free radicals pulled electrons clean off other atoms, turning new atoms into free radicals, passing on the problem and beginning a chain of 'hunt the electron' within cells. Eventually, Harman imagined, such a rogue free radical might interact with an atom that sits in a particularly vital molecule – a molecule with an important cellular function, say. He imagined that if this were to happen too often, cells would hit problems. Ruin enough of these vital molecules with free radicals and your cell ages, Harman hypothesised. He set to work looking for evidence to support his hypothesis.

Harman experimentally observed the impact of free radicals on short-lived mice dosed up with radiation. He found that after loading radiated lab mice up with antioxidants, they fared better – they lived, on average, longer. It wasn't as simple as he imagined, though. In Harman's tests, he was unable to alter their maximum lifespan, but he could experimentally increase the average life expectancy by up to 45 per cent, mainly through the use of the antioxidant butylated hydroxytoluene (BHT), a molecule widely used in the oil industry to prevent free radicals from oxidising fuel. Science operates most efficiently when someone comes out of nowhere with big ideas, and Harman was that scientist. Like white blood cells surrounding a cocky invading bacterium, academics soon surrounded him, probing and pushing around his ideas, trying their best to squash them where they lay.

But what they found was that Harman's hypothesis was interesting enough to be considered: free radicals did somehow appear to be involved in cellular ageing. Though researchers exercise extreme caution when it comes to making general statements about the relationship between cell death and free radicals, many have come to be swayed by the evidence found to support the relationship. If you slow the free radical damage you can slow the ageing. That's Harman's legacy.

So let us return to Ming and animals like Ming. Do ocean quahogs somehow tame their free radicals? The answer is yes, somehow they do. By comparing polar bivalves* with temperate ones (whose lives are often shorter), Eva Philipp, a molecular cell biologist at the University of Kiel, reported in 2007 that polar bivalves 'not only show lower metabolism but also show lower free radical generation, higher antioxidant capacities and in line with this a slower decline in mitochondrial function and accumulation of oxidative damage products with age'. They've mastered the art of free radical wrangling, in other words. And they aren't the only ones, it seems. There are many other wranglers of free radicals out there, which we are only now coming to understand. The birds, particularly, are striking examples.

Rather than root themselves to the seafloor and do very little, like clams, birds manage to live a life of extreme metabolism, and they manage it for years longer than many mammals could ever dream. There is a chance that I have watched puffins the same age as me (34), which is impressive enough, since puffins are actually quite small, but there is a wild Laysan albatross that is still pumping out

*Bivalves can be loosely described as molluscs that live their lives within two hinged shells. They include oysters, scallops, clams, cockles and mussels. Many species burrow but some, like clams and mussels, are adapted for life (as adults) on rocks near the shore or on the ocean bottom.

chicks on Midway Atoll at the tender age of 63. The oldest bird in captivity, Cookie (a Major Mitchell's cockatoo at Brookfield Zoo, Illinois), is still going at 83 years old. This is incredible, by Planet Earth's standards. A lucky pigeon may live 10 or 15 years whereas a similar-sized mammal – say, a rat – would be lucky to live three. Birds really are an anomaly.

Hummingbirds are a particularly striking example. A tiny hummingbird's heart pumps up to 1,000 times a minute as the bird zooms from flower to flower to flower. They are one of most intense metabolisers in nature. You'd think they could only last weeks at such a pace but, incredibly, they can survive for 10 years, consuming an estimated 500,000 litres of oxygen (per kilogram) during this time. On paper, with so much metabolising going on they should be riddled with rogue free radicals, but … they're not. Or they don't seem to be. Nick Lane's book *Oxygen: The molecule that made the world* does the mathematics far better than I could: based on a hummingbird's oxygen consumption and its longevity, the average bird's exposure to free radicals should be 10 times that of a similar-sized mammal like a rat (and maybe twice that of a human). Surely, they should be accumulating toxic levels of free radicals?

Gustavo Barja, a biologist at the Complutense University in Madrid, Spain, and his team have built upon Harman's early work on free radicals using birds as their model. By investigating pigeons, Barja confirmed that the mitochondria from pigeon tissue ate up three times as much oxygen as rat mitochondria. Yet the amount of free radicals produced in these reactions was far less. In fact he found that pigeons produced only 10 per cent of the free radicals that rats produced. Just 10 per cent. How they do this is still anyone's guess. And they aren't the only ones, like clams, to have somehow made free radicals impotent, robbing them of their destructive powers. Somehow, mechanisms to deal with free radicals have evolved in many animals that natural selection has thrown into positions of power, or at least niches like flight, where predators are almost non-existent or are stopped, on the whole, in their tracks. Bivalves, opossums, some turtles and some tortoises – all of them may, to varying degrees, have pulled from out of the evolutionary toolbox a probable wielding of antioxidants to allow for a longer lifespan. Why? And how?

Scientists (now thousands of them) are making great strides to answer this question. Flight appears to be particularly associated with the taming of free radicals because bats also manage, somehow, to control their release. Another link that highlights the relationships between flight and free-radical taming is that some non-flying birds, like ostriches, no longer appear to control their free radicals in the same way. They become more like ground-dwelling mammals of similar size (though they still fare better than most non-birds: a healthy ostrich might live 40 years).

There are yet other creatures that appear to tinker with how they deal with free radicals, and one of them may offer us a striking method through which we might modify our own longevity in future. The larva of the pearl mussel, *Margaritifera margaritifera*, is a particularly interesting example. It's a parasite that lives for part of its early life within the gills of Atlantic salmon, where it requires an extra year of the salmon's life to continue its own full development. There are early indications to suggest that the pearl mussel larva might somehow introduce a peptide into the fish's cells that mops up free radicals to ensure the fish achieves longer life, thereby allowing the tiny pearl mussel larva to complete its own life cycle and survive to adulthood. The parasite wants the fish to live longer. Fish infected with *Margaritifera margaritifera* are more resistant to tumours, and their wounds heal more quickly too. Though the exact mechanisms through which this tiny mollusc achieves longer life in the salmon aren't totally understood, there will no doubt be some scientists who wonder whether this little larva could be engineered so that it could attach to a human, an idea I rather like. It'd be like the Babel fish in *The Hitchhiker's Guide to the Galaxy*: just inhale some pearl mussel larvae up into your lungs every now and then and – rejoice – enjoy 40 more years of healthy life! The idea sounds ridiculous, of course. But then, why not? Not for the first time in this book, I find myself wondering about how much I would pay for such a treatment. Indeed, I think I would probably pay at least £50 a month to rent a pearl mussel larva …

At this point, I sound like I'm saying free radicals explain ageing and that this theory of why and how we age explains everything. But it doesn't. Not totally. For there is evidence to the contrary. For instance, in 2009 experiments with roundworms indicated that hobbling the production of naturally occurring antioxidants (which mop up free radicals) actually increased lifespan rather than decreased it. And then there are naked mole rats. They really are strange in all sorts of ways, and it's with them we should finish this chapter because … well, they shoot the free-radical hypothesis to pieces.

When German naturalist and explorer Eduard Rüppell first discovered the naked mole rat he assumed it was somehow diseased, so strange do they look. Hairless, aged, twitchy. Raging. They hang upon life's tree as if held by a novelty fridge magnet that no one can remember being given. They look totally out of place on Earth. But of course, they're not. Natural selection makes places, after all, and there are niches for those that can fill them underground, providing animals don't mind that they will probably end up being chiselled by natural selection into strange and (to some) rather ugly shapes (some scientists call naked mole rats 'penises with teeth') and sexual arrangements. Like ants and termites, naked mole rats have evolved a complex social arrangement, where underground colonies are populated by workers that support a single breeding queen who (literally) bullies them into subservience. If you care to look at a naked mole rat individual you will notice that it looks haggard and aged, and this is partly because many of them are, yes, haggard and aged. They live a long time. Where a similar-sized mammal may be lucky to live three or four or five years, it was discovered (through captive collections) that naked mole rats manage six times that. Many naked mole rats end up approaching 30 years, which is almost Ming-like, comparatively. Because of this, like pigeons and hummingbirds, they have become a model animal for those interested in age research, aided by the fact that they are relatively easy to keep in lab conditions, and relatively easy to monitor.

In environments with fewer predators (such as in flying vertebrates, or for shelled invertebrates like quahogs) natural selection appears drawn toward producing bodies that invest more heavily in parts and maintenance. Naked mole rats have become almost like high-mileage concept cars in their levels of maintenance. They are, like Ming, outliers. But the more we learn about naked mole rats the weirder it gets. Firstly, their cells appear to be ravaged by oxidative damage from free radicals, yet the strange thing is that it doesn't seem to bother them in any real way. Research suggests that their lipids, proteins and DNA are up to eight times more damaged by free radicals than in mice. Yet they appear totally fine with this. They continue as normal. The second thing that's weird about naked mole rats is that their cells appear to contain a host of strange proteins, which somehow assist other proteins to keep active and in good condition. And the third thing is this: naked mole rats don't appear to suffer from cancer. They just don't get it. This one is particularly weird because nearly all animal life can suffer cancerous growths, and nearly all animals have weapons to keep cancers in check; weapons that are, most of the time, very effective. Somehow, naked mole rats have gone further. They've nailed cancer. Naked mole rats are foolproof. Survivors. Battle-axes. And no one knows yet how they do it. Maybe it's from them that we might better combat the illness that riddles the lives of so many that we know and love.

From Medawar to Harman; from bats to birds to mole rats and Ming – animal life is showing us the way. The way to live, if not forever, then for longer. The way to live healthier lives in which we might only visit hospitals a handful of times during our lives. But free radicals, interesting as they are, offer us only hints at what is going on. The real action happens in those parts of our cells where free radicals are made: the mitochondria. For it is in our mitochondria that the story of ageing begins and ends and where the action really hots up. It is no wonder that many scientists are now focusing on mitochondria when looking for answers. By incorporating them into our cells, we got the keys to the castle … if only we could understand the bloody locks.

CHAPTER FIVE

This is a Dead Frog

'THIS IS A DEAD FROG.' Alison was my deathsplainer and these were her words when we met. They were words which rang in my head as I arrived home from the north-west with the dead bird under my arm. I thought hard about her opinion that we need to talk more about dead things with children. My skin crawled that I had ever said to my own children that some dead things we came across 'were only sleeping'. Alison's voice echoed in and out of my inner monologue. 'We need to give them access to it,' she'd argued. 'We need to get them used to the idea of life and death.' I pulled the bag to my chest. Perhaps this dead bird would give me my opportunity, I thought. Perhaps this would be my time. My eldest daughter, Lettie, was then three and a half years old. She was approaching what I'd class as full human cognition, her neurons connecting in such a way that she was starting to understand that other people have ideas and emotions that were different to her own. Now was the perfect time, surely, to talk to her about death. This would be the moment.

I went into the backyard and placed Alison's bag on the patio floor, and beckoned her over. 'What's that?' she asked, playing in the garden. I opened the bag and began to explain. 'This is an animal that I have collected from a friend,' I said, speaking clearly and calmly. I pulled out the ice cream tub which had been wrapped up tightly in cellophane, with a further layer of kitchen towel over the top. I was crouched down on the floor, with her standing a few feet away looking slightly suspicious. I began to unwrap it. 'Erm …' she said. 'What … what is it?' I peeled off the layers of cellophane and gently reached into the tub to grasp the peaceful-looking body of the magpie in both hands. It felt surprisingly floppy when I first grasped it around the mid-body. Its head rolled back when I began to lift it out, giving its expression a kind of exasperated quality. It felt surprisingly warm, like there was still some life left in there, which I suppose there was (just not magpie life). It was beautiful, though. Really beautiful. The warm sunshine caught the reflective green character of its wings and neck perfectly. It looked composed in death. Arranged, somehow. The wonderful remains of this creature rested in the centre of my hands, its eyes closed, beak tight shut. There was no blood. It was a peaceful thing. This was it, I thought. Here we go. I looked my three-and-a-half-year-old in the eye. 'This is a dead bird,' I said brightly. She said nothing. I tried again: '*This is a dead bird.*' I said it slightly louder this time. Nothing. She looked at it without saying a word. She pondered though, which was a start. She looked quietly confused for a few seconds, as if expecting the bird to do something, and then smiled at working it all out. 'Ahh,' she said proudly. 'It's sleeping.' She put her finger to her mouth as if to tell me to be quiet lest we wake it up. 'No … no.' I looked her deep in the eyes once more. 'It'll never wake up,' I said. Lettie was silent. 'It lived for two or three years,' I said. 'But now it's stopped. It stopped living. It's dead.'

Oh shit. I immediately realised the mistake I'd made. Three-year-olds are *obsessed* with their age. Totally obsessed with it. Would she take this as my perverse way of telling her that she wouldn't live to see her fourth birthday? That she might die like the magpie after her three years of life? She said very little. She chose not to move away though, which was something. There's loads of things for three-year-olds to do: spill things on the sofa, run around, make a mess, not help clean up. She chose to do none of these things. She chose to stay with me and the dead bird. I took that as a good sign. I stroked it gently, running my forefinger down the back of its head and down its spine toward its long tail feathers. Still Lettie said very little. I must give this another go, I thought. I mustn't mess this up. We must talk more about death without me somehow scaring her. I encouraged her to come a bit closer. I stroked the magpie a bit more and beckoned her to do the same. It looked so clean. It was surprisingly calming. Lettie didn't want to stroke the dead magpie. She did look closely at the magpie's face, though. 'It's eyes are closed,' she said. 'It's sleeping,' she said again. I smiled patiently. 'No.' I calmly recited once again the rules: all animals live for a bit and eventually they die. 'It's not sleeping,' I said. 'It's dead. It has stopped living.' She slowly started to get it, I thought. 'Yes, it's dead,' she agreed a minute later. 'It got died,' she said. 'Dead,' I corrected quietly. Marvellous, I thought. We looked more closely at it. I pulled its wing gently out so we could marvel at this spectacle, a flying theropod dinosaur. I told her about the fate of the dinosaurs, how this magpie's ancestors survived whatever it was that killed the rest of the dinosaur family and how amazing that was. At this point I started, as happens a lot when I talk to my children about theropods, to lose her interest. She glazed over. It was time to stop. I carefully packed up the magpie back in its ice cream tub.

'Will the magpie still grow?' she asked later that day. I was glad she was still thinking about it. 'No, it won't,' I said clearly. 'Dead things do not grow.' She paused. 'It won't grow,' she said, imitating my solemnity. 'It won't grow because it's dead.' Brilliant, I thought, she's got it. 'And I won't grow either until I'm dead,' she added quietly. I paused. 'Wait … What?' I ran over to her. 'Lettie, no, you'll still grow. Lettie, you're not dead!' I said this to her with a deep seriousness. In fact, I didn't let her get into bed until I was sure that she grasped that she was alive and definitely not dead. Forty-five minutes later I turned off the light. This was going to be harder than I had anticipated.

I turned my attention now to the dead magpie and my original plan. The reason I had fetched it from Alison was to learn about how, and to what degree, other corvids (the family that includes crows, magpies, rooks, jays and ravens) might examine this dead bird. Would they view it as an interloper? Would they feed off it? Would they express stress or anxiety at the sight of it? Would they mourn or have a 'funeral' as some scientists believe that they do? Or would they completely ignore it? I knew that, probably, it would be the last, but still, I thought I'd give it a try, just for interest, really. I bought a camera-trap to record what would happen to the magpie after I left it out there in the field, exposed. But where to put it? Our small cottage sits within an enormous colony of jackdaws, and occasional gangs of magpies also make themselves known to us, cawing and clacking a passage through the trees under which we park our car. There is a small clump of trees with a bare mound beneath that forms a stage for their interactions, and that seemed the perfect place to leave the dead bird and prime up the camera.

I left the magpie there that night, set the camera-trap up and returned each day to collect the memory card and see what had happened over the course of each 24-hour period. I had never done anything like this before. It was actually very exciting. Each day I sprang out of bed eager to see what kind of soap opera might have been recorded on the previous day as these social birds hustled and bustled around the dead magpie, possibly mourning it (I allowed myself to imagine) in all sorts of strange and unique ways. At least, this is what I had hoped. In reality it didn't work out quite like that. It didn't work out that way at all. Each file on the memory card was a 10-second bit of video footage, and on the first day of recording I had nine files to watch. I had been very excited at this. Might I see a world-first? I wondered. Something amazing that could find its way into this book? I clicked on the video files. No. The answer was no. The first video clip showed a woodpigeon shuffling through the leaves, totally oblivious to the dead magpie. The second video clip showed a woodpigeon, the same one, I suppose, shuffling through the leaves pecking here and there at scraps. The third showed two woodpigeons. They ambled in opposing directions from the centre of the frame. Not once did they stumble across the magpie. In the fourth video literally nothing happened. I'm not sure what set the camera off here – probably a butterfly. The fifth was the same. Oh well, I thought, there's always tomorrow. In each video file, the magpie lay there like Juliet on a plinth awaiting Romeo, its feathers ruffling slightly in the breeze like a black shawl. It made a beautiful image in many ways: a sleeping woman in a black cloak, being slowly consumed by summer leaves gathering upon her.

In the following 24 hours the camera trap picked up more videos of woodpigeons. And there was a squirrel too, piddling around in search of nuts. Day 3 was better. On day 3 I caught the movement of a larger animal. It was a cat. In the early light of dawn, the eyes shone like fire in the infrared footage. First it looked at the camera suspiciously, then stealthily approached the magpie. It gave it a few sniffs before looking warily around. I wondered if it might play with it, or pull it about a bit, but it didn't. It took a few moments and then it surprised me: it curled its legs underneath its body and, quite surprisingly, lay down next to the bird. It blinked its eyes slowly. The cat looked very content. The footage ended. It might have had a little sleep. Hmmm, ok, I thought. The next video clip was in daylight, a few hours later. The magpie was still there and the same, very annoying, woodpigeon was wandering around in its usual annoying way, stumbling around like a dull-eyed teenager looking for a wallet among sofa cushions.

And then, on the fifth day, just as the magpie was starting to really smell and I was becoming worried what the neighbours might say, something happened. I visited the camera-trap as I had done every morning, and I found that the magpie was gone. Gone! I was delighted. Eagerly I pulled the memory card from out of the camera and headed back to the kitchen, hoping beyond hope that whatever happened to the dead magpie had been caught on camera. Lettie was having breakfast as I did this. I dumped my laptop onto the table beside her as she ate her cereal, and I inserted the memory card. There was only one video. Would it be the one?

Lettie finished her breakfast and came over to sit on my lap as the video loaded. Nothing much showed up at first. Silence. Just the magpie in its usual position. A few moments passed. And then, from deep out of shot, a fox cantered in, its nose pressed firmly to the floor like a bloodhound in an airport lounge. Then the fox stood stock-still, it's long body framed perfectly in the shot. It came upon the magpie. It looked haunting and spectacular in this well-framed shot; its tail bushy like a paintbrush, its eyes, as it looked at the camera, like lit torches. It approached the magpie, sniffed it a bit, and then, like a lion gently picking up a cub, it put its jaws around the magpie's midriff and lifted it gently up. I thought it would leave at this point, but it didn't; instead, it swivelled its body away from the camera and pressed its rump up to the lens. Its tail lifted up, it squatted slightly, and from its anus it issued forth an enormous dollop of faeces, right up close to the lens; it felt like it was coming through the fourth wall. Then, its load lightened and with bird in mouth, the fox trotted merrily out of view. The video clip stopped. Wow. I chuckled, finding the whole thing totally brilliant and just, well, wonderful. I noticed, though, that Lettie, on my lap, did not find it funny. She sat there, still. She had a face of icy concern. 'Was that … was that in our …' she quietly murmured. Tears pooled in her eyes. 'Was that wolf in our garden?' I gave her a warm cuddle. 'It's not a wolf! It's a fox,' I said breezily. 'Foxes are ok. They're lovely. They hoover dead stuff up. They help nature.' This didn't really help much. 'That was *our* bird!' she said, suddenly indignant.

I didn't really know what to say to that. It wasn't ever our bird. It didn't belong to anyone. There was a pause. Her sense of injustice seemed to pass and she went back to being scared again. Foxes have as bad a reputation as wolves in Britain. Every single cultural reference her little brain has received so far in her little life has painted foxes and wolves as monstrous. Foxes are bad guys in many of her books and TV shows: *Peter Rabbit*, *The Little Red Hen*, *Dora the Explorer* and *Br'er Fox*. And now here I was, showing her videos of the foxes that live just outside the back door, right outside her bedroom window, stealing things, like dead birds, that she held dear. She gave me a look that indicated that her whole worldview had changed for ever, and we should all give her a few days to catch up so don't rush her, ok. I chose not to tell my wife that it was all my fault. First there was the dead magpie. Now the death-eating fox. This was my second opportunity to talk about death with Lettie, but … I felt it didn't go well. I could have done better. I kept messing it up. If this had been a test and Alison were judging, I felt like I would have failed. I chose not to tell Alison any of this, of course.

Six months later I would have another opportunity to make things right. My third. And this really was the big one. We had unfortunately heard of the death of an elderly relative, Lettie's great-grandmother, and we were all very sad. Lettie had visited Great Grandma with us frequently, first in her flat in the East Midlands and then, as she deteriorated, in a nursing home. We broke the news to Lettie after dinner. 'It's sad,' I said as we prepared for bedtime. 'We're all very sad.' 'Yes,' she said. We'd warned her it was coming, so in many ways she was ready. She was certainly suitably morose when we told her but it was hard to know what she really felt about it. A small part of me wondered if she was acting out how we were behaving like it was some sort of game. When she spoke there was almost a theatrical lilt to it. 'Great Grandma has *dead*,' she said, quite seriously. She looked at the floor, a mournful look on her face. I coughed. 'Umm, *died*,' I corrected her. 'Great Grandma has *died*.' 'Yes, dead,' agreed Lettie. 'Yes, she is now dead,' I said sadly. As I put on her pyjamas she shook her head sorrowfully whilst looking at the floor. 'Yes, dead,' she said. 'She has *dead*.' I had to correct her about this. It seemed important that she get it right. I paused and drew breath. 'Great Grandma has *died*,' I responded. She looked at me with sad eyes. 'Yes, she is *died*,' she said, her eyes full of mourning. I paused and took a few deep breaths. 'Yes, she is *DEAD*,' I said. 'She did *DEAD*,' Lettie confirmed. We were speaking quite loudly now, her raising her voice to counter mine. I rubbed my eyebrows and composed myself. Another deep breath. 'Yes,' I said calmly. 'Yes,' I repeated. We can tackle all this later, I thought. I stopped talking for a while.

As I tucked her up for the night, I nudged Lettie one more time into death-chat, just in case there was anything else she wanted to say before bed, partly out of concern for her mental well-being but also because this had all become very interesting to me. 'I miss Great Grandma,' she said, arranging her teddies next to her pillow. 'We miss her too,' I said quietly. 'She died,' she said again sadly. I stroked her arm. 'I know, but she lives on ...' I said. 'Great Grandma lives on in our heads.' I tapped the side of my head, smiling. I gave her a kiss and walked out of the room. Glancing back as I turned off the lights I saw her face. Lettie was sitting bolt upright in her bed clasping her head in her hands, with a look of confused terror on her face. I had just told her that Great Grandma had died and was now living in her head. No wonder she was freaking the hell out. I quickly turned on the lights and rushed back in laughing insanely. 'She lives on in our *hearts*!' I said, my voice shrieking wildly. 'In our *HEARTS*!' I said. 'Sorry, darling! My mistake! I meant we remember her *in our hearts*!' I was almost screaming. 'OUR *HEARTS*!' I said again. Lettie gave me a look. She knew something weird had just happened. I feared at this moment that I might have just shattered the emotional development of my eldest child.

A few days passed, and we all talked more about death and the upcoming funeral quite a bit, so that helped. This funeral was to be a big hurdle in helping Lettie to understand death, I'd realised, but to be honest by now Lettie was becoming something of a death-pro. She had come on leaps and bounds each and every day since we'd taken to discussing it. Alison, my deathsplainer, would have been proud of me. Honestly. Things were going much better. Lettie knew the correct phraseology, the grammar; she could express sadness and say really surprisingly comforting things. We'd succeeded, I thought. At the funeral itself she was an exemplary mourner. She sat with head down throughout the whole thing. She sang '*All things bright and beautiful*' with the appropriate amount of sadness mixed with happiness at all the jolly bits. She put her hands together and looked like she was praying really hard too, again at all the right moments. She commented on the flowers and cuddled family members outside when it was over. An exemplary mourner. She really was. We all said so. We drove to the wake, which took place in a lounge outside Great Grandma's old flat in the sheltered accommodation where she had lived before the full-time care. Lettie played with the other kids at the wake while the adults all made small talk and caught up with one another and talked about Great Grandma's long and quite happy life.

Things took a turn later that afternoon. After about an hour Lettie came up to me requesting a visit to the toilet and it was then that she betrayed her true understanding of death. We had walked down the corridor to the visitor toilets and we had found that they were engaged. 'Hey Lettie,' I said. 'While we're waiting for the toilet to become free shall we go and walk down the corridor to see the front door of Great Grandma's old flat?' 'Yes!' she said firmly. 'I want to do that!' Then she stopped and thought it over again. Yes, she thought. She gave me a look that suggested to me that yes, definitely, this would be an appropriate thing to do. And so off we went. We walked down the long snaking corridors toward the door to Great Grandma's flat, which was now empty. And then, as we approached the door to her old place, I noticed I was now walking alone. Where was Lettie? I looked behind me and I couldn't see her. What had happened? Then she came around the corner. And she was creeping. She was tiptoeing down the corridor toward me. 'Lettie?' I asked, slightly nervously. 'Lettie, what on earth are you doing? What … what are you doing?' She looked at me with incredulity, surprised I'd even have the gall to ask. She put her index finger to her lips. 'Sshhhh, Daddy,' she said as she approached the front door. 'You'll *wake Great Grandma*!' she hissed.

Oh dear, I thought. So I had failed. No, actually, *she* had failed. Failed to grapple with what death is. But I had failed too. I had failed because I'd assumed too much. I had totally misinterpreted her understanding of death. I had projected my own experience of what it is to mourn onto her. Thank God Alison wasn't there, I thought. Alison would have her head in her hands at this.

It was deeply interesting how Lettie had us convinced, though. Convinced that she had understood death. She had convinced us that she knew Great Grandma was in the coffin. But she hadn't grasped even the most basic bit: that Great Grandma had gone and she was no longer asleep in her bed. That she had been burned in a coffin and now she didn't exist, apart from in a nebulous atomic form floating around in the sky, and in ashes. That she had gone. That she had died. That she was dead. I wondered about whether, even if I had seen corvids apparently show an interest in my dead magpie, I could ever be sure what they knew about death. I realised again how hard animals are to read. I couldn't even work out my own child, for goodness' sake. Our perceptions of death are complicated. I had a long way to go.

PART TWO

THE EXPERIMENTAL PIG PHASE

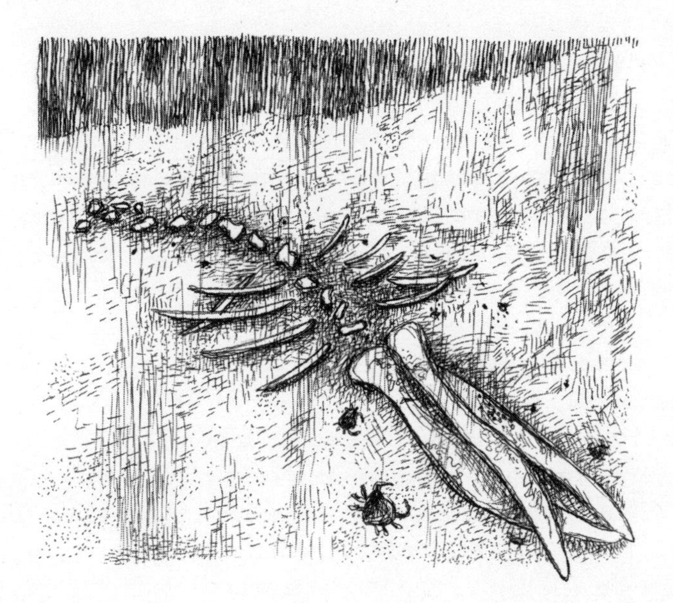

CHAPTER SIX

The Circus under the Tent

In life, it's the occasions where you have to wear waterproof trousers that are generally among the most interesting. Dashing through streams, pond-dipping, surveying rock-pools – warm memories are made of such experiences. But there are other reasons for wearing waterproof trousers. For instance, there are those rare occasions when you are told you may have to crouch down inside the collapsed carcass of a putrefying pig corpse. This, as it happens, is one of those days. 'If you're writing about death you have to go and see Peter Cross at TRACES,' was Alison Atkin's final piece of advice after handing me the dead bird. 'What's TRACES?' I had asked. 'Well, it's a bit like a body farm,' she said. I had heard of body farms on *CSI*. They have body farms in America. I had to go. Alison set it up.

I am in a 'secret location' near Preston in the north-west of England, and Peter Cross is my guide. As well as being an accredited forensic anthropologist, Peter is also a lecturer in forensic anthropology at the School of Forensic and Applied Sciences and he runs TRACES (Taphonomic Research in Anthropology: Centre for Experimental Studies), which was established by the University of Central Lancashire. Peter doesn't immediately strike me as someone at home in the lecture theatre. In our early emails he seemed more of a practitioner than a lecturer, somehow (though he later tells me he won 'Lecturer of the Year' in 2013 so … well, what do I know?). A previous meeting I had arranged with him had to be postponed after he was almost called to assist with forensic identification after a fatal plane crash in eastern Ukraine, close to the border with Russia. TRACES is largely the result of Peter's hard work. At its most simple, TRACES is essentially a large tract of land within which pigs are left to decompose under a range of conditions. The rates of decomposition for each variable are measured, creating a kind of death-measuring station – a 'rot-clock' – with potential application in human forensic cases. Say, for instance, a man's body turns up after months and months left in a waterlogged field; TRACES research on dead pigs left in wet fields can support forensics teams in their attempts to ascertain a potential time of death for the human remains.

I can safely say that I have never been to a research station like TRACES. I feel slightly out of my depth during my visit, and slightly insecure about the whole thing. So why am I going, you might wonder? I decided that it'd be useful to learn more about the ecosystems that emerge on the bodies of dead creatures, and Peter has kindly volunteered to be my guide, provided that I wear my own wellies, refrain from taking photos on my phone, and bring my own washable waterproof trousers. As we drive onto the site, Peter sets the scene: 'One of the key questions that's asked when a body is found is often "*When did this person die?*"' he says. 'What we focus on here is understanding the variables that influence the decomposition process and therefore may have implications for the accurate estimation of post-mortem interval.' He brakes suddenly and gets out of the Land Rover to open the gate to the site. Getting back in he adds quietly, 'That time period between life and death is much more tricky than *CSI* will lead you to believe.' There is a slight glint in his eye as he says this, so I sense that Peter doesn't like the hit TV show *CSI* – I make a mental note to avoid mentioning it or anything like it throughout our time together.

Today would be mostly about pigs, then, I think happily. In parts of North America, as many readers will be aware, forensic scientists use real human (donated) bodies for this kind of forensic research, placing them in institutions loosely called 'body farms'. At present this isn't the done thing in the UK, so we use pigs instead. 'Though it might be possible in the UK with a change in legislation …' says Peter. 'Well … it's not been done before here, and there are still mixed feelings about it within the anthropological community,' he explains. 'After all, is it ethical for us to study the decomposition of a human body?' I shrug my shoulders. There are benefits to using pigs, though. Peter prefers using pigs as human-models mainly for the strength of empirical data this opens up. 'Firstly, with pigs,' he says, 'you have more bodies to work with. In terms of replicate numbers, our experiments are … well, let's just say that they're more robust. We can have replicate groups of 20, 30, 40 – whatever we want. With pigs, we can test hypotheses very robustly, and this has real value.' I suspect this isn't the first time that he's talked to visitors about the relative merits of pigs. 'In this kind of research,' he says quite happily, 'there will always be a place for pigs.'

We park up. The TRACES lab itself is housed in a little courtyard at the entrance to the site. Next to the parking spaces there are some empty bathtubs dotted around, and some empty aquaria. There is a hose to hose everything off, and, I assume, to hose us off when we finish here. Peter walks toward the large metal outbuilding in which the lab is found. He beckons me in. Inside is a collection of tables, a kitchen, a couple of lab benches and an enormous tractor for carting the pigs around on site. The lab is well ordered. Everything has a place. There are hygiene posters on the walls, labelled drawers and hangers for lab coats and waterproofs and a footwell for the wellies. As Peter changes his shoes, I head over to a lab bench upon which sit a host of assorted pig bones. I scan the bones. The only ones I can immediately recognise are the vertebrae and, unsurprisingly, the ribs (which I have seen countless times on people's dinner plates). Then I notice a few more. There's a jawbone. A stocky femur. Scapula. There are many teeth. I try haplessly to plug them into a piece of lower jaw while Peter dons his waterproof trousers. The small talk continues between us: postgraduates, dissertations, academia, collections. Peter takes a bit of warming up, and I sense that, underneath, he might be a tiny bit worried I'm devising a trite 'UNDERCOVER CSI!' piece – a vainglorious article that sexes up the steely-eyed scientists who solve murders in their spare time by slotting together pig skeletons and lining up dental records and gunshot residues and the like. He doesn't know that at that exact moment I'm quietly pondering whether pigs lose their baby teeth like humans do (which, it turns out, they do).

'Ready?' says Peter suddenly. Already he is walking off into the site itself. I hurry behind him out of the lab and we make our way up a long farm path, lined on each side by thick scrub. It's a steep path and we climb upwards sharply, enough to make me a little out of breath. There is plenty of activity in the bushes as we stroll past. A pair of magpies monitor our passing. A mixed flock of great tits and blue tits moves above us, and some long-tailed tits flitter from one side of the path to the other, moving with purpose through the trees and bushes like a troop of investigative chimps. Peter opens a large gate in front of us and announces our arrival onto the site proper. An enormous flat expanse on the hillside stands in front of us, which is almost bowl-shaped in parts. There is a real nature-reserve look to it: managed mosaics of grassland and scrub (some succeeding into beech saplings and hawthorn) give way to an unmanaged area that is peppered with meadow flowers, and newly planted woodland borders about half of the site's 13 acres. A slight wind moves before us, grooming the grassland. It's a secluded spot, with only one entry point: perfect for studying what happens to dead pigs without worrying about what Peter calls 'human interference' (I don't ask what this means but I assume it is deeply frowned upon).

Looking across the flattest part of the site from the entrance, it's possible to see a series of wire-mesh wooden frames dotted around. These are placed in long rows and under each one are the remains of a dead pig, some of which have been here for more than three or four months, Peter tells me. The nearest wooden frame is about 20 metres away. We start walking toward it. 'At this point people often vomit,' says Peter deadly seriously as we trudge through the long grass toward the long-dead pig. As we get nearer and nearer I expect to be overcome by stench. But, no. It's actually fine. I manage ok. As we approach closer I expect to see some sort of body lying there. But, no. There's not even a proper body with mass or anything, just a collection of some kind of detritus lying on the floor. A messy collection of bits. And that's when it hits me. Only when we stand directly above the remains of the pig do I suddenly smell it; the stink wafts up in an invisible foggy cloud, entering my mouth, my nose and my pores. Staining my clothes. It's not disgusting, exactly – it's just ... powerful.

I don't feel nauseous or anything like that. In fact, if I were to describe it, I'd say that a dead pig, after three months in the sun, smells exactly like pork scratchings, the treats made from cooked pig skin and fat offered up in British pubs. The pork-scratching smell probably shouldn't have come as a big surprise to me; looking down upon the dead pig's body, long sheets of skin and fat are pretty much all that remains, baking in the sunshine. Lying there now, its body resembles a badly cut piece of grey carpet about the size of an ironing board. It looks nothing much like a pig. Long, thick hairs protrude from the carpet, I suppose, but I can't see any bones. No teeth.

I'm a little surprised at this, what turns out to be the first of many dead pigs that day. Coming upon it by accident, one might almost assume this was an oily puddle of thick wet rags, and little more. I put on my most professional voice. 'So … where's the head?' I ask, crouching down next to it. Now that he can see I haven't vomited, I like to think Peter warms to me a little from this point onwards. I like to think I have passed some sort of test. 'Well, there's a reason for that,' he says, crouching down next to me. 'The head is often one of the first places to go in pigs – and humans – because flies are initially attracted to natural orifices. They particularly like laying eggs in the eyes, the ears, the nose – as a result the head is often the first thing to skeletonise.' Orifices. Skeletonise. The conversation has escalated more quickly than I had expected it to. There is a hint of disgust that I do my best to mask. Peter pretends he doesn't see. He sees.

We look more closely at the pig. Its hide looks like plastic in the warm sun; there's a Vaseline-like sheen all over it, with yellow blobs that look a little as though Dijon mustard has been sprinkled in. This waxy substance, according to Peter, is called adipocere. It forms on bodies in wet environments, a leftover product of anaerobic bacteria that flourish upon fatty acids. For forensic scientists adipocere can cause problems, since it forms a boundary between the external environment and the internal environment – a boundary that decomposing agents struggle to cross, and which makes estimating decomposition rates slightly more difficult. Adipocere is a mummifying agent, essentially; some death professionals apparently call it 'mortuary wax'.

We sit there for a few moments, looking at the scene. A few bluebottles come and go, but there doesn't seem to be much going on. Until we look a little more closely. And then they start to appear: hundreds of tiny flies hover and perch upon the skin. They are tiny, like midges really. Some, if they land, run quickly up and down and across the surface of the pig, slaloming the thick hairs that protrude through the grey jelly of the adipocere. These are cheese flies: a worldwide family of flies, mostly about 3–4mm long. Up close they are almost metallic, and quite beautiful. A pair of prominent eyes take up most of the head; two wings are held politely over the back of the body. They have a kind of urgent, desperate grace to them. And the cheese flies are scavenging the corpse, as is their way. Two or three land and wander around to the underside of the mummified sheet out of our view, no doubt looking for darker places to lay eggs. Their grub-like larvae are often called cheese-skippers, for their apparent ability to fling their bodies forward through the air by holding onto the corpse with their grappling mouth-hooks and flipping their back end over their head. Cheese-skippers typically take three weeks to pupate, and are one of many species used by anatomists to determine how long a dead body has been lying there. They are one of hundreds of body clocks that apparently make deceased megafauna (I use the term loosely) their home.

I look a bit more closely still. There are other small flies, too; they dance in and out of the corpse. And tiny beetles. On closer inspection I see a host of green beetles flutter busily in and out, going into a hole that leads into the darkness under the skin. They look like tiny bees going in and out of a hive. I try to get a few photos of these electric-green beetles as they land, to no avail (Peter has let me take some photos on the proviso that I don't upload them onto Facebook; I imagine how many of my friends and family would reach for the 'Report Abuse' button if I ever posted pictures like these). Peter offers to lift the sheet of mummified pig flank so that we can see underneath and I stand back to give him some space. I ready myself for what lies beneath. Peter firmly grabs the pig skin at two ends and pulls it back with purpose, like he's pulling a tablecloth out from a perfectly laid table. There is a sudden gush of activity underneath, as if Peter has pulled the lid off some sort of tiny flea circus, rudely interrupted midway through a performance; a kind of instant chaos in miniature. An invertebrate melee. A host of small beetles retreat desperately under remaining flaps of skin, many appearing midway through copulation. Five or six larger beetles retreat into the surrounding long grass beneath our feet. A couple of micro-moths flitter past us, and cheese flies lazily buzz off to find other corpses to feast upon. There is much more to see with the skin pulled back, I realise. And it looks more like a pig now. The pig's teeth and jawbone are dotted around one end of the mass, while vertebrae are pulled out of place – ransacked, almost – and show remnants of once having been a chain (I would say they are now a loose arrangement). Rib bones on the floor separate the activity of the remaining tiny beetles into paddock-like enclosures.

This is all incredibly gross, but also wonderful in equal measure. In my years of turning over bits of wood looking at invertebrate life beneath I have never seen such … occupancy. In the centre of the chaos sits a football-sized mass of thick viscous yellow foam that looks like some sort of swirling primordial-ooze-cum-circus-tent. I try and make out whether the foaming ooze is the remains of the pig or an emergent property of the life that has since stirred within it. After a bit of scrutiny I am still none the wiser about this foamy blob. I get up close with my camera. It moves slightly. Peter reads my face. 'Ahhhhh, that ...' he says with a certain amount of ceremony, '... that is a maggot mass.' Ok, I think. Of course. Peter crouches down once more beside me. He looks closely at the foam. 'So ... hmmmm ... these look like first instars ... yep, so this is the second wave of maggots,' he says with sudden certainty, standing up. He looks at me. 'The second generation of maggots feed upon the tissues that remain in the chest cavity.' He stops. He reads my face. I hide a tiny little gag reflex. I try to pretend it wasn't a gag, but it was, and I have to wait for the feeling to pass. 'Wow ... thousands of them,' I say, trying to sound like a pro. But I'm a fraud. He knows I just retched. He knows I'm a fraud. Still, I tried. It was involuntary really. It was, after all, a reflex. I'm a little bit weird about maggots for reasons I have never been sure about. Embarrassed, I hear myself lay out my credentials to Peter: that I'm a zoology nut and this is my life and it's only maggots that do this to me. Maggots and spiders. Maggots and large spiders and engorged ticks. Peter looks momentarily unimpressed. He says nothing. The sight of those maggots really does disgust me, though.

I take a moment, a few seconds to myself, before getting up close to watch the mass more closely. I gather myself fully and crouch down. This time I feel better. It is *fascinating*. It really is. Each individual maggot is about the size of a nail-clipping, yet together they have formed an orgy of decomposition; a great swirling snarly mass of purpose. These, according to Peter, are bluebottle maggots. Laid in their thousands and protected in the carcass, there is so much to eat that they have no need to move far; each maggot becomes surrounded and supported from all sides by other maggots. They feed together, and this foamy mass is what results. I watch closely again. Imperceptibly, like a minute-hand moving, the ball of them almost seems to have a life of its own. Revolving. It sort of swivels like a gas giant. I don't expect you to be quite as enamoured as

I suddenly turned out to be but … you know, there was real beauty to it. Truly, it was beautiful for a few moments. The whole thing was. And then – woosh – I suddenly dry-retch again. Peter clears his throat and pretends he was looking the other way.

I'm amazed that the pig's skeleton hadn't looked at all as I had expected. I'd imagined it remaining articulated or looking somehow like a perfectly preserved museum skeleton. But it had looked … vandalised. 'You'd be surprised what a large mass of maggots can move,' says Peter, smiling. 'And why is it so foamy?' I ask. 'Well,' says Peter, 'when maggots are actively feeding they stir up the decomposition fluids with air and this forms a thick bubbly foam.' He sticks his fingers into the foam, pulling it apart like curtains, to reveal a fresh maze of cleaner-looking maggots writhing beneath. Peter tells me that, by working their way between sinews and joints, the maggots loosen the skeletal structure. Gravity then does the rest; the skeleton falls apart. I comment to Peter about how it looked like the maggot mass was swivelling. Though I'm sure it was a trick of perspective (or my head spinning), apparently, yes, maggot masses may in fact slowly rotate, though much more slowly than I could really have noticed. The phenomenon, Peter tells me, may ensure that heat is moved out of the centre of the mass, where unlucky 'inner maggots' may face conditions 20°C warmer than at the mass's edge. On the level of selfish genes, it may be that hot maggots head to the cooler edges, dragging cooler maggots into the fiery recesses below. Only further tests (involving the tracking of individually marked maggots, which is apparently very difficult) will reveal the inner physics of maggot masses.

I stare at the maggot mass in silence while Peter attends to something else. It hits me looking at the maggots that in a matter of weeks many of the thousands upon which we gaze will be fully fledged bluebottles, some of which will probably hang around and lay further eggs on this dead pig, and others nearby. Other bluebottles will migrate further, making their way into nearby people's houses or the McDonald's down the road; eaten by birds or hit by cars or fly swats. So many of this pig's atoms will migrate away within the adult bodies of the flies that leave. It seems obvious but … there it was.

We stroll off to see more of what Peter has started to call his 'experimental pigs'. There is a familiar chaos to each corpse that we look upon: the smell, the initial disarray of fleeing creatures, the vandalised bones, the maggot mass. But, like seeing a mass of crowded shoppers from above, order comes from it once you sit and wait and watch for long enough. The cheese flies. The tiny green beetles. The moths. As predictable as wildebeest, giraffes and elephants in their own way. There are the same water holes. Similar scaffolding. Hiding places. Places to bask. Places to bury. I later learn that the green beetles are one of many species of rove beetle. Looking a little like earwigs, 'rovers' have short forewings (or elytra) and long thin bodies that lack pincers. Though it doesn't look like it, they can apparently fly. They move from feeding station (in this case: dead thing) to feeding station (ditto) to live, breed and lay eggs. It turns out there are quite a few rove beetle species out there. In fact, there's rather a great deal of them. Apparently there are more rove beetle species than there are fish, mammals, amphibians and reptiles combined: more than 60,000 species at last count. In the US, one out of every five beetles is a rove beetle. You may know rove beetles – they can move their abdomens from side to side or up and down – a feature for secreting chemicals that aid defence, or perhaps for chemical mimicry. Some are beautiful, with elytra that are deep greens or metallic blues. The largest in Britain is the Devil's coach-horse (adults are about an inch long), which looks and acts very angry indeed when disturbed, but actually is mostly harmless. The rove beetles on the experimental pigs were far shinier and far more charismatic. I came to think of them as Day-Glo fruits of death. According to the books, rove beetles are among the second wave of invertebrates to arrive at a fresh corpse. They had been there quite some time.

It's probably time for a brief description of the specialist waves of invertebrate immigrants that move onto a creature like a pig once it dies. And it really is quite a cast. First to arrive are blowflies, such as bluebottles. As is well known, these are perhaps the most active and fastest colonisers of dead bodies on land. Hundreds may appear within hours. Thousands sometimes. They know death. They know its smell. And they smell it well (but more on this in a second). After this come the cheese flies, the flesh flies (or onion-flies) and the house flies, all of which sow their genetic oats on the fertile soil of death. Some beetle species also turn up early to the feast. The carrion beetles (often called sextons – the Nicrophorinae) are one such. These beetles actively feed on the corpse both as adults and larvae (which, apparently, consume regurgitated food from their parents). As a result, many of these carrion beetles actively compete with blowfly maggots for food; they undergo a race against time to reach a fresh corpse first. Natural selection – which finds a new gear when intense competition erupts between individuals or species – has thus provided many carrion beetles with an incredible ability for chemoreception. They are in an evolutionary arms race against the blowflies; a kind of mad sprint to the death exists between the two groups. Flying through the air, these beetles use their sense of smell to locate hydrogen sulphide, a common chemical released by bacteria in great clouds during decomposition. The same is true of the blowflies with which they compete. Incredibly, some carrion beetles have pulled in hired help: they carry around with them a tiny mite species that devours blowfly eggs – a handy tool for carrion beetles seeking to eliminate any competition that may exist at a new corpse.

After the blowflies, the cheese flies and the sextons arrive, then predictably the things that eat the blowflies, the cheese flies and the sextons put in an appearance. This is where many of the rove beetles fit into the picture. At TRACES I saw them amongst the bloated remains of the pigs picking off blowfly eggs and larvae, carrying them off victoriously in their jaws to hidden depths beneath the flaps of skin. It was like watching from a helicopter as a tiny leopard pulls a tiny antelope into a tiny tree to eat it. But it's not just rove beetles that hunt within the corpse. As decomposition continues and the body dries out a little, other predators turn up. Small chequered beetles (with their long, armoured bodies) may make an appearance. They too are drawn to the broiling maggot mass, coming for the good times and staying for the hospitality that a dead pig provides to their offspring. Rotund clown beetles (great name) may also turn up, again drawn to the feast. Sepsids (often called black scavenger flies) make occasional forays into and around dead mammalian bodies, though we saw none on the day at TRACES. Sepsids apparently sit on grass perches next to the corpse, waving their wings to one another suggestively, enticing others closer, looking for opportunities to mate. Some parasitic wasps will visit the site, undertaking aerial raids on the maggots, immobilising them one by one and laying their eggs within their paralysed bodies (the going rate for some wasp species is 12 offspring per maggot, which seems like a healthy return). Dung beetles too may visit – drawn to the rotting intestines of herbivorous mammals particularly. They work on the raw ingredients of the faeces because, well, why not?

Soon the entire corpse is in the 'advanced decay stage'. Life is everywhere at this point – hundreds of individuals of hundreds of species, and that's not even including the bacteria and fungi that thrive in newly flourishing niches. The role of bacteria is particularly vital in decomposition, for they break down the complex molecules of life into individual elements, like carbon, nitrogen and sulphur, which can then be absorbed by plants and fungi. It really is a wonderful and life-rich place. A food web has emerged from the chaos that looks as rich and full-bodied and enigmatic as any other place on Earth (it is just that those interested in studying it may need to possess a nose and stomach like Peter's).

But, even on a dead pig, there are still other creatures yet to join such a circus. Skin beetles, with oval-shaped bodies covered in tiny scales, are among the most numerous. Their larvae make short work of the dry scraps left on bones and sinews, and are particularly useful in cleaning skeletal museum specimens before public airing. After the skin beetles come the clothes moths (your suits, dresses and jackets are a pale imitation of what their larvae have evolved to devour). But there is more still to come. As the season changes, the corpse is no longer simply a source of food: the remaining bones and sheets of adipocere become a place on which new creatures can climb, and in which shelter and dry egg-laying sites abound. By now the circus has largely come and gone, but the circus scaffolding remains and it has its uses. In come the marauders: the harvestman arachnids, the centipedes and the larger ground beetles, as well as different species of rove beetles, searching for spoils and other late-comers to the party.

These pigs, their dead bodies; it really was a weird experience. A wonderful gathering of the unusual, the monstrous and the bizarre – all of which have made their living for millions of generations from recycling the atoms of the once-living back into the living again. In front of my eyes, watching those pigs, I was watching nature denature itself and spontaneously renature itself into something else. And this is what has happened to almost every animal that has ever lived (until recently, when humans opted out by choosing cremation, selfishly starving many thousands of beautiful rove beetles in the process).

Surprisingly few animals have had their post-life fauna and flora identified, and the theatrical cast that colonises a given dead thing probably differs a little bit (or quite a lot in some cases) for each class of animal that dies. There is diversity among corpses. Among the best documented accounts of animal decomposition are those of marine megafauna, mainly whales, whose bodies drift deep down onto the seafloor post-death. What happens to the enormous bodies of whales is not so different from that which occurs upon the bodies of dead experimental pigs, though there are no insects and there is the added possibility of sharks, which livens most zoological accounts up immensely. No wonder it is a well-told story. Still, I think it deserves its place in this book because it is … well, it is, like the pigs, extraordinary.

There, at the bottom of the sea, away from the influence of weathering and where shallow scavengers rarely penetrate, whale remains can provide the energetic needs for other creatures not for years, but whole decades. Some of the best-studied 'whale falls' (as they're called) are from North America's Santa Cruz Basin. Blurry camera footage shows dead whales on the seafloor 2km below the surface. In the videos (all available online for those interested) their enormous bodies lie there, lit by submersibles that look like space probes. The whales' enormous eyes are glazed over, their great mouths gape. Without life, the whale communicates in a new language: a song to scavengers. As decomposing bacteria flourish within its body, smells are released. These chemical signatures drift through the deep ocean currents, sideways and upwards. They contain messages. Messages that other life will soon intercept and interpret.

Over the coming days and weeks the hagfish are among the first to pick up these messages. Thousands of them undulate and ripple across the surface of the whale like ribbons, making its whole surface appear to writhe. Lacking a true backbone, hagfish represent a distant and rather primitive fish – they have no jaws, only simple eyespots and no true fins. However, to call them basic would be to do hagfish a disservice. Being an early representative of life after the split between vertebrates and invertebrates has its perks, after all, for hagfish possess the best bits of both taxonomic groups. Being worm-like, they are adept at wriggling into the occasional orifice. They can also, if the need takes them, absorb dissolved organic matter across the skin and gills. Like leeches they can go months and months without feeding, drawn toward food via sets of special chemoreceptors that work in ways we can only imagine. Like slugs and snails they can slime up for defence and, like insects, they possess antennae (barbels), tactile organs that come out from the mouth. Yet this is, by all accounts, a fish. A jawless fish. We think of worms and fish being distant; in the hagfish, their shared history is exposed. They are without doubt one of the most under-celebrated forms of life on Earth and they're down there as you read this, wreaking seven shades of hell upon the dead bodies of every whale lucky enough to have escaped the whaler's harpoon in the night.

But it isn't just hagfish. There are other creatures heading the dead whale's way. Along with the rippling hordes comes another fish with its flippers firmly in the past: the sleeper shark. A shark in all but the eyes, which have a zombie-like tone to them – sunken almost into the toothy jaw. Sleeper sharks look very stoned almost all of the time. They slowly plunge their jaws into the bleached whale flesh and twist their heads in sluggish jerks that tear it from the body in plugs, which they then swallow whole. Rattail fish (more properly called grenadiers) – ghoul-faced deep-sea fish – also make an appearance. As do the shrimp-like amphipods and crabs. Predictably, it isn't long until the zombie-worms (great name #2) join the party.

During this early stage of decomposition, 60kg of meat may be consumed each day by the denizens that have made the dead whale their new home. Many will have sex in and around the whale, too. Many, many, many creatures make use of this time of local bounty. Time passes, though.

The whale begins to change. After months of activity its boat-sized body begins to fall apart and scatter itself; its bones become exposed, and organs protrude and rupture, spilling more organic matter onto the seafloor. New real estate is born during this period. New exploitative organisms appear – they are the enrichment opportunists, and among them are snails, worms and bacteria. Amongst the worms are the snotworms (great name #3) and bone-eating bristly marine worms (#4), which are part of the polychaete family. They squeeze into the whale's bones and, with the help of bacteria, break down and digest the fats and proteins present (how they find a dead whale in the first place is still anyone's guess, but it's likely they arrive as swimming larvae). Over decades these worms, working in tandem with bacteria, devour the enormous bones almost totally, whilst other bacteria swarm in great mats over their surface. And it is upon these bacterial mats that other animals then gather. Mussels, clams, limpets and sea snails; they feast for decades, possibly a hundred years or more, on these bacterial mats. One animal has made a reef. It all started with one whale.

These strange ecosystems at the bottom of the deep seas have probably been present for hundreds of millions of years. Where now such animals flourish upon the bones of whales they once would have fed on extinct marine reptiles like plesiosaurs, mosasaurs and ichthyosaurs. If whales understand death, one presumes they'd be flattered by the life they can sustain for so many, for so long, after their passing. But alas, it is only sperm whales and beaked whales (which make regular trips down to such depths) that may ever know what awaits them. Perhaps they are filled with horror when they come upon such a scene on the ocean floor? Perhaps they gather together and mourn? Or dance? Or tell stories? Unlikely, but still, we may never know.

After two hours of walking the site with Peter I became no more hardened to the sight of entrails and drying pig flesh than I was at first. I certainly didn't gag again, mind.

But I definitely found myself close to gagging a lot of the time. By the end of my visit to TRACES the smell of pork scratchings was ingrained so deeply within my lungs that I was worried I might be a health risk to family members later that night. My clothes stank of it. Totally stank. 'Have you got used to the smell?' I had asked Peter at one point. 'Doesn't really bother me,' he said. 'I still smell it, I register when it's strong but …' He shrugged a little. 'I lived on a farm when I was younger and every job I've ever had to do was in some way related to death. I've smelt it all of my life.' Graduating from farm work, Peter worked for 15 years in the abattoir industry, enforcing regulations that ensure meat is prepared safely and to certain standards. After veterinary public health he moved into forensic anthropology, working over a number of years investigating mass graves in Guatemala and, later, in New York doing some death-scene investigation work and undertaking more forensic anthropology during time with the New York Office of Chief Medical Examiner. After New York, Peter came back to Britain where his research interest in death further blossomed, and his dream of a research school like TRACES were made real.

What I particularly liked about Peter was the respect he had shown whenever the conversation had turned to human death and human suffering. As we had stood in front of a decomposing pig hanging from a tree (to mimic a human hanging victim) he had talked with sadness about the human hangings that he had tended to. He talked almost with his head down, suddenly becoming quite shy and almost mournful. He was sombre whenever real human death was mentioned. At no point did he play up the harsh brutality of his work. At no point did he shoot for the ghoulish or say anything macabre. He wasn't trying to impress me in any way. This was his work: death was his science. I realised now why he probably didn't like the whole *CSI* thing, and I'm glad that I had never really thought to bring it up. I had noticed the same serious tone in Alison's voice when she had talked about the victims of plague that have become her study area. Even though hundreds of years had passed, and these people were now just bones and not much else, her science was never cold or uncaring. Human remains were still human, like her. Human remains; the bit that remained. Her job was about people. It was about lives. Peter's and Alison's areas of work are not easy. Occasionally they must come across scenes that must be intensely emotional; scenes of past murder, suffering, even genocide and disease. I think the world is made a better place through jobs like theirs. I respect them both a great deal.

On my return to the real world many people asked how the trip to see the pigs went. 'What was it *LIKE*?' they asked. 'What did it *smell* like? What did they look like?' For a couple of weeks I was like a soldier coming home from leave being surrounded by civvies with a perverse interest in war stories. I held back the gory bits. I tried to be like Alison and Peter: rational and just, well, just … cool. But still I thought about it. I was particularly ashamed that I had gagged upon seeing the maggot foam. I was ashamed that Peter had seen. Ashamed of myself. The truth was I really had been momentarily disgusted by what I had seen. It felt involuntary. It had been out of my hands, really; I had no choice over it. It just kept happening. The maggots, I don't know … they unnerved me. It was the first time I've ever felt disgust in such a visceral way. I was momentarily intrigued about this feeling – in fact I thought about it a lot.

Mostly though, my memories of those pigs were about the amount of new life I saw. It was the beetles that I remember most fondly and which captivated me so. So much variety. So much diversity. I could wash off the smell of pork scratchings, sure, but the beetles remained burned into my memories and my experience of the day. The circus under the tent. The comings and goings. That each and every one had descended from generations of individuals that had made another creature's death their own precious precursor to life. God may have an inordinate fondness for beetles (so the old quote goes), but the enormous diversity of rove beetles betrays the mind of any potential Creator as having an inordinate fondness for death. I had loved it. But, still … I had gagged. The experimental pigs had produced within me a strange mix of feelings.

CHAPTER SEVEN

Sex and Death: The Contract Killer

It was my first real paid job in wildlife conservation. It had been offered to me so that I could contribute a small amount of my time to scientific research into frogs and their diseases, on one proviso: that I manage the Frog Helpline. This was the very early 2000s and most of the British population had no access to the internet in any way, shape or form. These people needed help with their frogs. These people needed help with their frogs urgently. These people needed a Frog Helpline. I was to become one of a handful of people in the history of the world to be able to offer free amphibian counselling over the phone. It was simultaneously the most brilliant and the most awful job in the world.

That's unfair. On the whole, I liked speaking to people about the relationships they had garnered with their backyard frogs, newts and toads – albeit I probably did this in an accidentally mawkish, patronising way. But I learned a key thing from this job, for it was the first time I appreciated the scale of death in nature. And this death was primarily frog-related. Each year we had hundreds of calls about dead frogs to the helpline, and they often came in spring. There were moments when some callers had cried with sadness at the annual loss of life in 'their' frogs during breeding time. 'There are at least 20 dead frogs!' they'd shout down the line. 'They're littering the pond! What is happening? What do I do?' they would shriek. What are you supposed to say in situations like that? I had received little or no help or advice about what to say about this from my supervisors. No one had thought to give me training for this sort of thing, so I'd tell them … I don't know … to net the dead frogs out of the pond safely and put them in the compost, I guess.

Most, after some counselling, accepted these strange incidences of frog death as another mystical quirk of amphibian-kind: hopping, breathing through their skins, singing, sometimes changing colour, sometimes poisonous and … sometimes, you know, dying all at once. It just seemed to add up … Frogs really are weird; this was really just another way in which frogs were weird. Many of the callers who expressed deep sadness to me were worried that the sudden mass death of frogs was their doing somehow. That they were to blame. They had been racked with guilt that if they'd changed the water or not cleaned out the pond the previous year it would all have been ok and the mass frog deaths would never have happened. I told them it wasn't their fault. Sometimes this helped; sometimes the caller was not convinced. Some people refused to take on board my assertion that this was totally normal and that many animals died *en masse* after reproduction. 'NO!' they would argue. 'It *had* to be disease,' they'd say. 'There had to be a reason!' Sometimes this would result in soggy packages being sent through the post with my name on them, the contents betrayed by being frog-shaped and stinking to high heaven. The words on the front of these envelopes were always scrawled excitedly with something like 'JULES HOWARD: A SAMPLE!!!' as if it would somehow make my day to receive something like a dead frog in the post and perform an autopsy on the mail-room table (even a book about life and death is no place to describe to you the horrors that lay within such packages). The truth about frogs really is simple, though. There is no mystery. Well, there is almost no mystery, I should say. Allow me to explain.

In very simple terms, it goes like this. Frogs have a contract with death, the same as anything else. This contract stipulates that they are going to die. Whatever happens, it's simple: they will die. So the frogs have a choice in what they do next. They can 'choose' to invest everything in one massive bout of sex right now (what I will call for the purposes of this chapter TYPE 1 FROGS), which will result in them almost certainly dying from exhaustion. Or they can 'choose' to have quite a bit of sex now and try to maintain their body for another year for more sex the following spring (the TYPE 2 FROGS). For frogs, this is a tricky choice to weigh up. The fact that there are a great many things out there that cause frogs to die – dry weather, disease, predators – means, inevitably, that natural selection is often drawn toward the TYPE 1 FROGS; hey presto, the world fills up with frogs who aren't too fussy about dying midway through (or hopefully just after) sex. That's economics. That's asset management. That's frogs. Of course, there isn't really 'choice' (at least in frogs), there's simply variation upon which natural selection has acted, is acting, and will always act. Frogs really do have it tough, but they make the best of it through their sex lives. And there really are many, many ways for frogs to die. I heard some very strange stories when I manned the helpline. Cases of exploding amphibians were one such weird one. It was a phenomenon I would come to hear about fairly routinely from 2005 onwards, thanks to one unusual news story.

'Toads in an area of northern Germany are being killed off by a mysterious disease – they are exploding,' reported *BBC News* rather breathlessly that year. 'Thousands of the amphibians have died in recent days in a pond in Hamburg's Altona district, with their bodies swelling to bursting point.' According to locals, at least 1,000 toads had died in this manner in a matter of days. All of them apparently swelled up to three and a half times their normal size and then simply exploded where they stood. What could cause such a thing? No one knew. Specimens were taken. Lab tests were inconclusive. There was no viral infection. No signs of a fungal infection either. Rumours flew around (bizarrely) that the phenomenon was linked somehow to some local racehorses kept in a field nearby, though no one seemed exactly sure how. Inexplicably, others talked of toad suicide, thankfully falling short of suggesting it had been some sort of ritual or pact. After a few days of this speculation German authorities chose to close the site, and the pond was inevitably dubbed 'The Pond of Death' in the popular press.

It took a few weeks for people to realise what was going on. In fact, it all came down to one man's relentless enthusiasm for solving amphibian mysteries. After examining living and dead specimens local expert Frank Mutschmann had noticed something strange about the dead amphibians. The toads' livers had been removed, he observed; apparently pulled through a circular puncture wound on the belly. 'There were no bite or scratch marks, so we knew the toads weren't being attacked by a raccoon or rat, which would have also eaten the entire toad,' Mutschmann told *The Independent* in May 2005. 'It was clearly the work of crows, which are clever enough to know the toad's skin is toxic and realise the liver is the only part worth eating.' Case closed. Except … for what reason did the toads appear to explode? The answer to this part of the mystery comes down to the common toad's impressive threat display. When exposed to predators common toads quickly open their nostrils and fill their bodies with air, blowing themselves up rather like a balloon. The strategy works well (particularly against grass snakes) but it is less effective if you have a hole in your belly through which other organs, like intestines, may inevitably become extruded with force. Presumably the local crows had stumbled across this delicacy and, being wily and inquisitive, the organ-harvesting practice had spread. To all intents and purposes, the following morning the toads did, in fact, look to human observers like they had exploded. But they hadn't. They had simply been operated on by the crows.

What's most surprising is that it ever became international news at all – exploding toads were by no means a new thing. In fact the 'exploding toads' phenomenon had been recorded in Germany in 1968, as well as in Belgium, Denmark and America in the years since. And I came across a few incidents of exploding toads on my days working the helpline, and never thought too much about them. It was nature. Normal. Normal-ish. Throughout much of the temperate world, spring is the only time of year when frogs and toads congregate, so naturally there will often be predators queueing up for a go. Herons getting rich on the stuff. Owls making ponds part of their night-time raids. Cats driven out of their minds with the profligacy of all those amphibians. Out-of-control Jack Russells in country parks; crows, magpies, jackdaws – the lot. Otters provide us with further interesting behaviour when it comes to eating toads. Sometimes I would receive reports of otters deftly collecting breeding toads in their hundreds, carefully biting the (non-toxic) legs off each, and depositing the bodies into a little pile next to the stream or river. Some otters were apparently known to peel the toxic skin off the toads before feasting on the skeleton and internal organs; they too were said to leave the skins in a little pile.

Passers-by coming across sights like these the following morning were often horrified. But a zoologist stumbling upon such a scene might have a different perspective – the message reads loud and clear to them. They might think: 'The chances of death here are quite high: these animals are probably going to be quite sexed up.' As I learned from my enquirers to the frog helpline, amphibian life really is hard. In environments of death, like those that occur during monstrous amphibian breeding congregations, animals have evolved to get their sex in relatively quickly. In Darwinian terms, it's better to have loved and lost than never to have loved at all. Common toads and frogs aren't truly semelparous though – they don't breed once and then die like Pacific salmon and mayflies – but many species like frogs are close to it. They sometimes dabble in it, let's say. In niches where death is likely, or an almost clear certainty, natural selection is drawn to the last throw of the dice. But it isn't always as simple as that. Death sometimes plays its card in hidden and secretive ways. Semelparity isn't always what it seems.

Take the example of Pacific salmon, for instance. Sure, swimming upstream through a gauntlet of predators like wolves and bears certainly would increase one's chances of death a great deal, but actually it may be that semelparity evolved in these salmon for another reason. It seems that salmon invest heavily in egg size, since larger eggs produce stronger fish fry which have a higher chance of survival when migrating back to the ocean to mature into adults. With a finite amount of energy in the pot, female salmon must fuel this egg investment with something. It may be that female Pacific salmon pay for this investment in eggs with their lives. Or it might be a mixture of the two things. Either way, the unshakable contract with death stands … even if the contractual wording is in a rather hard-to-read font.

A classic case of semelparity occurs in a small family of mouse-like marsupials, the antechinus. And again, the contract has come to take a strange and unusual form. What natural selection has done to male antechinus is almost grotesque. In nearly all species, male antechinus invest everything at the end of each year in one massive testosterone-filled sexual bender. Males attempt to have sex with every female toward which their legs can carry them. During this time the male's body gives up totally on maintenance. Males become riddled with gangrene as they quest for females. Their hair falls out in great tufts. But this doesn't stop them. Still the males go for it. Their immune system almost completely ceases to be, but still they rampage. Even with their insides bleeding and many of their vital organs totally corrupted, the males still show an interest in sex. After days and weeks of this, many males actually become a disease risk to females. They become (almost) the walking dead. Still they go on with the sex, though. Still they go. And then finally they succumb, all of them, to the degeneration of everything else in their body. It's over for them, but hopefully not for their offspring: the genes for semelparity persist. But from where did antechinus get their semelparous ways? What does their contract with death look like?

There are three hypotheses to explain male semelparity in antechinus. The popular view is that the mass death of males somehow means that the following generation is blessed with extra food resources (the contract being that males commit to death but succeed in securing a good inheritance package for the kids). Then there is the second view: that female survival is unpredictable, so males evolve to get it where and when they can (males commit to death but hopefully manage to have sex with a female that survives). And then there is the third view, which offers neither the selflessness nor the machismo of either of the first two hypotheses. The third hypothesis is that antechinus evolved semelparity not because of predators but because the females made them do it. And it is from Ed Yong's superb blog *Not Exactly Rocket Science* that I discovered more about this intriguing hypothesis.

In 2013, scientists (led by Diana Fisher at the University of Queensland) looked at the lifestyles of the 12 antechinus species and investigated other insect-eating marsupials, 52 species in total. What they noticed was that the further away from the equator they searched, and the more seasonally abundant the habitats became, the more likely it was that males would die after breeding. On the whole, species with 'suicidal males' were more common in seasonal climates where insect populations flourish in summer and then die back in winter, and suicidal males were less likely in climates where insects persist in good numbers all year round. Fisher and her colleagues think that as the ancestral antechinus species moved south from the equator, their breeding windows evolved and changed – rather than breeding all year, their breeding season became primed to coincide with the seasonal glut of insects. In other words, their breeding windows shrank enormously. This observation gave rise to Fisher's own hypothesis: the third hypothesis. The third hypothesis to explain semelparity in antechinus males is that they have been squeezed by the females' much shorter reproductive windows. All of their reproductive potential must be met in a matter of days and weeks, rather than months. The competition for sex is therefore rather intense. So intense, in fact, that it has become better for males to die trying than not to try at all. This is sperm competition at its most extreme. For these males, sexing-to-death is the only option left. The only game in town. Semelparity probably occurs in antechinus and not, say, in rodents due to the fact that antechinus are constrained somewhat by their marsupial ancestry. Where mice and rats can have multiple litters during the glut times, marsupials are stuck with a life-history plan that sees them give birth to tiny larvae-like young that require plenty of nurturing. In fact, four months' worth of nurturing. Males really do only have a limited window in which to act. Their contract with death offers little by way of wriggle room.

This small clutch of marsupials are the only mammals that actively have sex until they die. The only ones. And that makes them inherently interesting in all sorts of other ways. For instance, why is this behaviour so rare in mammals? And why is it so rare in any vertebrates, for that matter? Scientists continue to debate the many evolutionary drivers chugging away behind the scenes of semelparity, and the truth appears often quite hard to unpick, at least currently. The second question is particularly interesting, though. Why is semelparity so rare in vertebrates? The only land-dwelling vertebrates that exhibit total semelparity (apart from antechinus) are some of the *Hyla* frogs (including, fittingly, the gladiator frog) plus a handful of lizard species. And that's it. So why is it so rare? Perhaps vertebrate bodies, like German automobiles, cost a lot to make. Perhaps it's easier to keep them ticking over than to scrap them and buy another? We don't know. It's still a mystery. Invertebrates, it seems, are the opposite to us in this respect. It seems that invertebrates may be very cheap to make, so semelparity appears in many insects (particularly butterflies, cicadas and mayflies) and spiders and molluscs as well (including some squid and octopus species). Plants too, on the whole, appear to be quite cheap to replace. In fact, many species of plant favour semelparity. Some bamboo species, after decades of waiting it out, are known to suddenly flower and then die *en masse*, all at once (this has been known to cause local hell for panda populations as their required food source suddenly begins to wither and die in front of their eyes). Most grain crops and domestic vegetables are also semelparous. Their sex lives are influenced in direct response to the likelihood of dying that exists out there in the real world.

Death really is a contract, and animals have evolved to exploit its loopholes in the way that best suits them and the spreading of their genes. On the whole, though, you either throw down your hand and bet big, or bet small and hope to visit some other tables. Either way the house will always win in the end. That's the contract. Nearly all of the animals I had come to meet and would come to observe on this journey into the murky netherworld of life adhere to this contractual agreement with death. Nearly all of them share this basic principle. They will all die, sure, but in many cases their bodies have evolved to make the best of it; often to get rich or die trying. One assumes it will be the same on all planets that may or may not harbour life.

It has been more than 10 years since I worked on the frog helpline, but I still think about it often. It was my first regular experience of discussing animal death in such zoological terms. I felt really sad for the people who had come across such scenes in their gardens. Often they really were very upset by all the death, and I am not sure if I convinced many of them that it was all natural and totally fine, and that it wasn't their fault. I guess I will never know. I can't help but feel like I was communicating it all wrong, though. I wondered if I could have sounded more rational and scientific when discussing death, without sounding cruel or blasé or insensitive to their disgust and apparent dismay and sadness. The truth was I didn't have a clue. I was as muddled and confused about death as the public who were ringing in. I still am. Death seems like the most rational thing in the world. Yet we find it so hard to grasp sometimes. But that, in itself, is very interesting, I think.

CHAPTER EIGHT

Coffee with the Widow-maker

The images of the foamy maggot mass stayed with me for ages. It wasn't the maggots themselves. It wasn't the maggots *individually*. It wasn't the bones or the skin. It was … I don't know. It was the *mass*. The foamy mass of them. It brought on disgust. Total disgust. Are all humans genetically primed for disgust? I wondered. Is disgust an evolved avoidance technique to keep us away from the things that might kill us, as is often said, like maggots and snakes and things (perhaps like caterpillars) that look a bit like maggots? My mind slowed. Questions coagulated over the weeks that followed. Did the ermine caterpillars have it coming? I thought. Were they killed off because they reminded us of death? Were they killed off because of our disgust? I couldn't shake these thoughts. And then I had another encounter with disgust. A few months later I was pulled into the world of spiders for an unrelated project, and I got to thinking about spiders. Was it the same with them? I wondered. Could our disgust with spiders be equally evolved? A lucky break came my way and through this lucky break I got to see spiders and maggots from a totally new, and slightly unnerving, perspective. This is the story of that encounter.

In northern temperate countries, it's often September when the vast bulk of spider species are at their largest and most obvious. Hence, each year, in the UK at least, there is always a run of ridiculous scare stories – reports of spiders as big as dinner plates, spiders finding their way into people's mouths or ear-holes or nostrils. In recent years, though, these stories have taken a new twist. A single new spider species has come to steal the column inches normally devoted to such stories, and it has even stolen the front page on occasion. It's the noble false widow spider, a venomous spider (ok, someone needs to say it: all spiders are venomous) apparently capable of killing and maiming, even though the reality is that it's about as dangerous as a needle and thread left on a kitchen sideboard with no one in the room. But that doesn't stop it being a story, of course. In recent years, the press have absolutely gone to town on this poor spider. They've absolutely milked it for all it's worth. In the UK we have very little in the way of venomous creatures so this might be a perspective thing, I don't know. Either way, the noble false widow has become THE THING TO FEAR in modern times. The thing to hate. I don't like this kind of misplaced hate. I never have. It feels dirty and mean-spirited and ill-informed. Because of this, in the last two years I have been fairly vocal about them, patiently informing the public that the threat posed by the false widow spider really is minimal and trying to instil a love (or at least an acceptance) of spiders in the hearts and minds of the British public.

To be honest, I don't particularly like spiders (in fact, I'm actually a little phobic of them), but … I don't know … it's a principle thing, I guess. I don't like to see humans express such vocal hate, even if it is only toward spiders. This year, after a false widow piece I wrote for *The Guardian*, something incredible happened. I was contacted. Contacted out of the blue. I was contacted by a man. The man through whom it had begun; the journalist who had launched a thousand headlines by first running with the stories about false widow spiders. The man responsible for all of the public's recent fear and terror. 'I'm the one that started it all,' he had messaged me. Really? Wow! I'd written back to him quickly: 'It'd be great to meet!' 'Sure,' he wrote back. And so it happened. I was going to get a face-to-face meeting with a man I'm going to here refer to as John. John: the journalist who had dented the zeitgeist in such a way that many people woke up each morning thinking only of the threat that everyday spiders could pose to them, making everyone incredibly jittery and nervous. John said he'd prefer to remain anonymous if that was ok, in case his former employers heard he'd been spilling trade secrets. No problem, I'd responded. We set up the rendezvous – a coffee shop in east London. I looked forward to it.

Everyone in Britain right now is jittery and nervous about false widow spiders. Everyone in Britain now knows about them. Everyone knows someone who's seen one. Everyone is jittery and nervous and it's almost totally all John's fault. I was a tiny bit confused about why John was happy to want to meet me in person. In fact, why had he felt the need to contact me in the first place? What had urged him to say hello? I was his enemy, wasn't I? I wanted good things for spiders, didn't I? Perhaps he wanted to right an old wrong, I thought. Perhaps this would be his redemption. Was he racked with guilt at the fear and panic he had caused in the general public? Was this his confession? Truth be told, I had been worried that he might somehow want to trick me into exposing something new about spiders. A factual titbit that might get him another million hits. But I had got it wrong. When we met, John was actually lovely. A really nice guy. I had half expected him to be toothy, rat-like and fiercely mouthy but, sat next to me in the coffee shop, he was nothing but open, friendly and warm. There was nothing particularly shady about him at all. He wore baseball shoes like mine and a denim hoody, and in a nice little Sainsbury's bag he put on the table he had some flowers for his mum. Looking at him, he actually came across as almost a bit shy; really not as I had imagined at all. In fact, as we made small talk I realised we shared some things in common, both working in industries where the work is sparse, both juggling work with having young children. This wasn't how this meeting was supposed to go. I had imagined it'd be like the coffee scene in *Heat* between Al Pacino and Robert De Niro – him the gangster cooking up lies as part of a racket, me a loyal defender of the arachnids he aimed to discredit – but no, it was more … it was more Central Perk than that.

The public obsession that John had started surrounded the noble false widow spider – also known as the rabbit hutch spider – a non-native species that has been living in Britain for more than a century. This fairly innocuous-looking spider – smaller than the garden spiders that frequent our window sills and doorways – has certainly spread in geographical terms in recent decades (our houses are the most wonderful caves, after all). And yes, the noble false widow can just about bite if mishandled: their fangs are capable of penetrating skin, opening up one's body to secondary infections, some of which, if you are desperately unlucky, can kill. So yes, there is that.

The spider is capable of causing death in much the same way as an abcess can become excruciating painful and, if left unclean, can lead to serious infection. A threat, sure … but statistically, a highly improbable one (readers of this book from other countries will no doubt delight in the fact that this spider really is about as dangerous as an unwashed sewing needle and that us Brits are the most pathetic wimps when it comes to venomous creatures). The blind terror that this humble little spider can elicit in British people has reached epidemic status. And John, looking at me and smiling sweetly, must surely shoulder some, if not all, of the blame.

'The way it started was interesting,' John said, drumming his fingers on his coffee cup. 'I was at work a couple of years ago and I was interested in reading about spiders that are found in east London.' He sipped his coffee and paused for effect, almost like it really was a confession. 'I was just flicking through stuff online really,' he said. 'And then I came across a tiny story – a really tiny, tiny little story – about the false widow spider and the fact it had made its way to Essex. They were on the increase a little bit. And that was it …' He looked up at me. 'That was my story,' he said. 'I wrote the story. And that was it, really.' He shrugged slightly. 'Really?' I asked. 'It was that simple?' He nodded. 'I wrote a little piece for the paper's website – nothing much, just a few paragraphs,' John said. 'I showed it to my editor and he read it. He liked it. My editor looked at me: "*Seriously, go to town on this. Go to town,*" he said. So I did.' John smiled. The memory warmed him. 'Straight away, the story topped all other news on the website – and that was only referring to east London. We started wondering … how well would it go if we went national?' He sipped his coffee again. 'I spoke to my editor again. "*Well, let's broaden it,*" my editor said. "*Let's say these spiders are on the rampage across the entire country!*" So I did.' '*Rampage*?' I heard myself say quietly. 'I talked to the graphics guy,' continued John. 'I got him to do a map showing the spiders to be all over the place across the entire country. And that was it.' That was it. The story flew. 'False widow spider on rampage in Britain' said the headline. The subhead read: 'A DEADLY spider that can kill humans with a single bite is on the rampage in Britain.' We were off. The story flew across news channels, online, Facebook, Twitter, broadsheets, red-tops … all of them bought into it.

'But … wait,' I said, trying not to sound confrontational (which is easy because I'm not at all confrontational). '*Killer spiders*?' I tilted my head. 'They don't kill people!' 'I know they don't *kill* anyone,' said John. 'We all knew they don't *kill* anyone, but the way we saw it was, well, if you were allergic or something, you could die. They *could* … kill you … technically.' There was a pause. John looked at me for reassurance. I pulled a face that was somewhere between a smile and a wince, which seemed to resolve the situation. 'So anyway,' he went on, 'from that point onwards, they became KILLER SPIDERS! *MILLIONS OF THEM! ALL OVER THE PLACE!*' He flashed his palms beside his face in mock terror. The thing is, he wasn't kidding. The story really did get everywhere from that point. It blew up totally. It was the *Daily Star*'s number one story for many weeks. So what happened next was actually very predictable. 'So …' said John, '… obviously, my editor wanted to replicate that story. It was inevitable, I guess.' And so they did replicate it. And then some.

'FALSE WIDOW ALERT: MILLIONS OF KILLER SPIDERS ON LOOSE IN UK' (8th October 2013).

'FALSE WIDOW ATE MY LEG!' (10th October 2013).

'MUM'S TERROR AS FIFTY FALSE WIDOW SPIDERS RACE TOWARDS HER!' (13th October 2013).

'KILLER FALSE WIDOW SPIDERS ATTACKED MY LITTLE GIRL!' (17th October 2013).

'FALSE WIDOW SPIDER BITE LEAVES GRANDFATHER FIGHTING FOR LIFE' (23rd October 2013).

'We started doing more of them,' said John. 'And then others started picking up on the false widow story: the *Mirror*, the *Daily Mail* … even Sky News joined in.' I asked John whether, as a journalist, it was annoying when others leap aboard a journalistic gravy train like this. 'It was annoying,' he agreed. 'But it was also quite flattering to know that I started it, you know?' Again, I pulled my strange nod-smile-wince. There was another brief pause. 'The more we did it,' John continued, 'the more I started noticing that the spiders were making the front page of the paper quite a lot. It was …' John paused again, momentarily debating whether or not to tell me something I might find particularly insightful. 'Go on,' I smiled. 'Well,' said John, 'I got told that by putting spiders on the cover of the paper we would achieve an eleven per cent spike in sales.' 'Eleven per cent?' I said. I could barely contain myself. 'Wow! Eleven per cent!' We both said 'eleven per cent' a few more times. My jaw had dropped. 'Eleven per cent,' he replied. At seeing my reaction he seemed unable to gauge what face he should now pull. He was half very proud and half very ashamed. Perhaps mostly proud, I think. 'You look at the newspaper's web stats for the year and, of the top 10 highest-grossing stories, I think three of them were about spiders.' I was staggered that there was such an appetite for scare stories out there. Totally amazed. We stared out of the window in silence watching the busy street outside. If stories like this can raise sales by 11 per cent, I'm surprised there aren't more spider stories, I thought. Perhaps the noble false widow spiders had been lucky to have remained relatively hidden on these shores for so long …

I had expected John to take it all as a bit of fun; to pooh-pooh the punch of his stories as trivial and meaningless. But I could tell that he understood that his stories really had impacted on society. Something really did happen as a result of these pieces. It took place on 22nd October 2013, after a busy week of false widow spider stories, when it was announced that a school in the Forest of Dean, on the borderland between Wales and England, was to shut down temporarily because the geographically widespread false widow spiders had been discovered (apparently) on site. A letter to parents was immediately sent out. No doubt about it: with identification of at least one false widow on the site, there was a chance of spider bite. The school had to shut for two days. The spiders had to go. Thousands (probably hundreds of thousands) of invertebrates living on the site were fumigated, all because of a potential spider that posed about as much threat as a cactus (hell, the fumes left from the fumigation were probably more threatening to pupils). And it was all John's fault. Sure, no one had *died* because of John's stories. But hundreds of pupils had lost out on their education. Hundreds of parents had to face the unbelievable annoyance of spontaneously having to arrange childcare (which is unbelievably annoying). And it was all John's fault. Many say that such scare stories are harmless, that they don't have a noticeable impact and that there's no conservation significance if only a few more spiders get stamped on. But there was a real human impact from this story. Two days' learning lost. Panic. Ill-informed hysteria about a threat so tiny that it makes wasps and bees look like genocidal maniacs.

I brought the story of the school up with John. I asked him if he felt guilty about it. 'Yeah, a little bit,' he said. 'It was kind of weird, yeah – I felt guilty …' He paused. 'But then … I said to you just now that I started the false widow story, but perhaps it would have started anyway? Someone would have come across it,' he said. 'Surely with time, someone would have come across these spiders and written something similar.' He looked at his coffee cup. I could tell that there was another confession coming up. He paused again. 'When the paper started going overboard with it, I sort of decided I didn't want to run it any more. We did too many,' he said. 'We milked it.' He put on a mock voice and flapped his arms a bit: '*FALSE WIDOW SPIDER ATE MY LEG!*' he said. '*SPIDER ATE MY FOOT! MY HAND FELL OFF!*' We laughed. 'In the end the same stories were everywhere. Too many other papers were using it. It all got a bit … *samey* … in the end.' He sighed. 'I did feel guilty about it, though. I did.' He looked up at me earnestly. He considered his words a little before continuing. 'In the end,' he said, 'in a weird sort of way, my integrity was at stake. I'm not going to win a Pulitzer any time soon, but this isn't exactly the kind of story that I set out to write. I wanted to write the truth but … I guess I'm a bit of a web-traffic whore too …' He smiled. 'We all are,' I said. He nodded. 'I stopped doing them in the end,' he said quietly. 'It was time to stop.' John has no doubt that others will continue his legacy, and I fully believe him. In the coming years I suspect that we will hear more about the noble false widow. Others will carry on the legacy that John started, for the simple reason that fear and spiders and people improve sales by 11 per cent.

As we packed up our things I asked John why the stories had done so well. Why did it fly? John was instantaneous in his response. 'People like a villain,' he said almost matter-of-factly. 'If you can take a story and make it about a villain, especially if it's potentially life-threatening, then people listen.' To John, the spiders were destined to end up on the front page in the same way that if Peter Benchley hadn't written *Jaws*, someone else would have. Stories like these are inevitable, I guess.

I drove home, my head swirling with thoughts. At times we have a strange mutualistic relationship, journalists like John and I. He needs loud-mouthed science-literate people like me to get his quotes and give his stories more authority, but I need people like him to write scary stories so that I can write comment pieces about why spiders aren't scary and we should all love spiders a bit more. No media outlet would let me write about spiders without a hook like the one he provides. It's strange. What can I say? It's deeply messed up but, well, this is the strange loop in which many naturalists and zoologists nowadays find themselves. A web, really. A trap. Spin. The spider's trade and our own have become one. We need one another. But the school closure had resonated deep within me. I didn't like it. A thousand young people's education had been affected, a thousand families. I didn't like it one bit. 'Though it sounds remarkably improbable, we may have to accept that nature occasionally has the audacity to encroach fleetingly into our lives,' I'd written in *The Guardian* a few weeks previously. 'We may have to accept that, no matter how hard we try, we can't sterilise every single aspect of our lives from spiders.' But the evidence was there, plain to see, in how that school had reacted. We had already started trying to sterilise everything. We wanted them dead, just like we had wanted the ermine caterpillars dead. Was it because they were a genuine threat to life, or was it something deeper?

The meeting with John had come at an opportune time. I was still frustrated at Birchwood council's response to the ermine caterpillars. Do spiders and maggot-like caterpillars deserve such treatment? Is it really in our genes, as we are so often told, to hate such creatures? Is it natural for us to be scared of animals that we associate with death? I think possibly, but probably not.

Darwin was among the first to notice the universal human response to disgust. In *The Expression of the Emotions in Man and Animals* (1872) he wrote:

> *With respect to the face, moderate disgust is exhibited in various ways; by the mouth being widely opened, as if to let an offensive morsel drop out; by spitting; by blowing out of the protruded lips; or by a sound as of clearing the throat. Such guttural sounds are written* ach *or* ugh; *and their utterance is sometimes accompanied by a shudder, the arms being pressed close to the sides and the shoulders raised in the same manner as when horror is experienced.*

Darwin was basically right. Type 'disgusted face' into Google and you'll see for yourself: row upon row of faces of all ethnicities with the heads pulled back into the shoulders, their upper lips pulled up to their nostrils, a kind of snarl almost, with their eyelids half closed. Often the face is pulled to one side. It's the classic URGH face (I'm doing it right now, for effect). It's easy to label this as a clear evolved reaction: a simple instinctual response to limit our contact with potential pathogens, pulling away our various orifices and brain extensions (eyes) from the objects likely to kill us. But this could also be a 'just-so' story – things look fit for a purpose, so they probably are.

The truth is that it may be more complicated than this. Much more complicated. Smell appears to be a key conduit to instigating a human disgust face. Particularly the smells of death, it seems. The two smells that waft most freely from the dead (in plants or animals) are cadaverine and putrescine. These two simple molecules are released in great clouds that waft upwards as key proteins in decaying bodies are split apart and break down. Put simply, the more protein an animal has, the more putrescine and cadaverine molecules are produced and the more potent we find the smell. Cadaverine and putrescine have proven rather a handy scent for animals to become sensitive to, as I learned by spending time with the dead pigs. Scavengers are attracted to them, for instance, while many things positively avoid them. What is particularly interesting is that fish may have broadly similar systems for cadaverine and putrescine detection to what we have, suggesting a deep history of death-aversion inherited (possibly) from shared ancestors. Key olfactory receptors are often called TAARs (trace amine-associated receptors). In zebrafish, for instance, some TAARs latch onto cadaverine molecules and send electrical signals to the brain, which the fish register as odour. They work very much like our own receptors, which are also TAARs. Indeed, similar TAARs-like systems probably occur in invertebrates too, not least blowflies, who use their own sensory equipment to locate (rather than flee from) dead bodies. Such insights as this really do hint at the ancient origins of such a signalling system. All of us have probably found death unappetising for millions of years, though we may never have consciously understood why until this point in time.

That we have sensory equipment that detects the smells of death is pretty much what one would predict from natural selection: that the key equipment to notice things associated with death, and respond accordingly, has survival value. What appears less clear is what exactly it is about death that drives this evolved response. Is it because dead bodies are a pathogen risk to animals? Or is it because avoiding dead bodies means that living animals avoid getting into similar scrapes, whether it be contact with a predator or any other unknown threat such as disease? For some animals it could be either or it could be both. We will probably never know. The fact that many animals respond similarly to the smell of faeces (another pathogen breeding ground) may provide a clue, though. Spoiled food, faeces, death, disease – they're all sides of the same square: in other words, evolved recognition software that says AVOID WHERE POSSIBLE. Faeces-avoiding behaviour has been observed in sheep, cows and horses and probably also serves to reduce the chances of such animals accidentally gobbling up parasitic worm larvae, which often live in faeces. It's said that many primates (and wild reindeer) also display such faeces-avoidance behaviours, led largely by smell. Death – or rather parasites – is possibly why.

Could human disgust really be an evolved human universal? Professor Paul Rozin of the University of Pennsylvania is today's leading proponent of the idea, arguing that the response keeps us from ingesting items that may be riddled with pathogens, such as blood, decaying meat, faeces or vomit. If Rozin's hypothesis is right, it gets me off the hook for retching upon seeing my first maggot mass. If he's wrong, I guess I must hang my head in shame: I am simply a massive wuss. If Rozin does prove to be right, then perhaps the public's response to the ermines was ok – the ermines really did have it coming for behaving like writhing maggots dripping in great silken stalactites on the passers-by below. And it really was very maggot-like. So was their destruction simply a case of mistaken identity? It will forever remain a mystery, probably.

And what about the spiders? I thought. Were we born to hate spiders? According to the press, yes. This popular notion has developed in recent decades that humans have indeed evolved to fear them. Journalists are so quick to believe this that they no longer check if it's really true. But have we? Is it in our genes? Have we *really* evolved to hate spiders? In reality, no one can be sure, so perhaps I can offer up my two cents on the subject.

There is a weakness in the central pillar of this idea about us having evolved to hate spiders, and that is simply that human cultures around the world – unlike with disgust – appear to vary in their responses to spiders. Not all cultures immediately display a disgust face. Not all of them shiver and are terrified. And not all humans are scared of spiders. In fact, strangely, it appears to me that what we see with spiders appears to be the *opposite* of what we might expect through natural selection. One would predict that adverse behavioural responses to spiders would be most prevalent in the ecological zones in which biting venomous spiders are most common. Yet this isn't what we see. Britain has no venomous spiders directly capable of killing, yet we stand on chairs and many of us do seven shades of hell's worth of screaming upon seeing one. Folk from other countries (particularly Australia – a nation that has, on the whole, learned to live alongside venomous creatures) must laugh at us. And there are other observations that don't seem to fit the facts. You'd expect, for instance, children, being smaller and more susceptible to death by envenomation, to be especially wary of spiders. But no. That's not what I see. In my experience of working with thousands of young people, often watching and sometimes handling spiders, they start off loving them until about age four or five or six, when suddenly their attitudes change. They begin to run a mile at the sight of a spider, sometimes only at the mention.

And there are even more problems with this received wisdom about spiders being a thing that we have evolved to detest. For instance, other animals are far more threatening than spiders and we certainly don't display the same disgust or fear about them. Mosquitoes (through malaria) are a killer of more than one million people each year and evidence suggests that we have been troubled by the malaria parasite for at least 80,000 years (probably longer). If we have evolved to fear spiders, you'd think we'd evolve to fear mosquitoes too, right? But no. The sound of a mosquito buzzing past one's ear remains little more than an annoyance at best. Few of us wet the bed with terror upon hearing one. Few of us get on a chair and become paralysed with fear. (I sound like I'm belittling the torture and agony of those who are phobic. This is not my intention. I am actually a bit phobic of spiders myself, remember? But I think that I got my phobias from awful childhood experiences with spiders, *not* from my genes. And as an aside: I find clowns and collections of holes* scary too. No one would ever argue that this is genetic.) So if I'm right and it's not genetic, then what is it with spiders? Why did John hit the mother lode when writing about his false widows? What part of the cultural funny-bone was he tweaking?

North America has its own version of the false widow spider, called the brown recluse spider. The popular notion

*This is a genuine phobia called trypophobia. Before you ask, I have no idea what childhood experience gave me a fear of lots of holes.

is that this little spider is rampaging state by state across the continent, lurking in dark corners waiting to bite. As with the false widow, the popular press seems to get itself into a whirl whenever (rare) brown recluse bites result in (even rarer) disgusting secondary infections that require surgery or, worse, prosthetics. Like the noble false widow, the brown recluse has captured the American public's imagination. Many Americans are, it seems to me, quite terrified of them. People report bites and skin lesions from all over North America, even though the species only lives in a thin line of states between the Rocky Mountains and the Appalachians. There are the same scare stories too. The same images of necrotic infection, blamed on the spiders, that we see in the UK. People *know* what they saw: what they saw was a brown recluse. Similarly here, people *know* what they saw; what they saw was a noble false widow. But the reality is that spiders – all over the world – are incredibly hard to identify and tell apart, many even by experts. Species are often misidentified, and these misidentifications often unnecessarily cause fear and panic and terror.

Consider this case study: in 2005, the arachnologist Rick Vetter (from the University of California, Riverside) asked the American public to send in specimens that they suspected to be brown recluse spiders. Of 1,773 specimens sent to Vetter from 49 states, less than 20 per cent were actually brown recluses. In subsequent similar research, discussed in *Wired*, Vetter showed that similar misidentifications had occurred by those who called themselves entomologists, physicians and (worryingly) pest control operators (the cynic in me calls foul play). Like the false widow, the brown recluse spider has hardly crawled from the depths of hell. The biggest specimens only reach 20mm for starters. In fact, the brown recluse can barely bite at all, its fangs being hardly able to penetrate human skin. They certainly don't hunt or stalk humans. Far from it. In the spider's home range, humans and brown recluses probably live in something close to harmony, where neither much bothers the other. In one extreme example, a 2002 survey by Vetter of a 19th-century-built occupied home in Lenexa, Kansas, found 2,055 brown recluses living with a family, 400 of which were likely to have been big enough to cause envenomation. No one had been bitten in that house, ever.

In fact, there are many interesting similarities between the brown recluse and the false widow spider, but most interesting is how they have captured the media interest. Our ears prick up when hearing negative stories about them. We love it. 'The press has, by and large, painted spiders in a negative light,' says entomologist Chris Buddle of McGill University in *Wired*. 'People jump at the chance to hate spiders. It's easier to vilify them than to adore their biology and natural history.' We think nothing of the days and days and days and days when nothing bites us or bites those that we love. Yet as soon as we hear a story about spiders it is as if our worst fears have been realised and confirmed: spiders are out there, ready to bite, and we've been lucky to make it this far alive. We tell as many people as we can. We chatter. We talk. We share. We are eager to spread the word about encounters with spiders, even if the word is hokum. Are we the only animals to do this? For a few weeks after meeting John, I wondered if there were any parallels with other social animals in the animal kingdom when coming across potential agents of death. And then I came across one. A study involving, once again, those most intelligent birds, the corvids.

Do any non-humans get a kind of hysteria about novel threats to life? Yes, it seems that some might. This is a study about fear from a zoological perspective. A study that displays perfectly, I believe, why we all get so het up about spiders. And it really is a simple and very cleverly designed experiment. It was a 2011 study undertaken by scientists at the University of Washington, led by John Marzluff (who co-wrote *Gifts of the Crow*, mentioned again in Chapter 15), which looked at how American crows responded when faced with novel and unusual threats. They did this, simply, by becoming the threat. Researchers trapped, banded and released 15 or so crows before releasing them unharmed, and they wore a mask whilst doing this. A rubber caveman mask. The crows, naturally, didn't like to be banded. They made a mental note: *the human with the caveman mask is a total bastard. Avoid the human with the caveman mask.*

But what the researchers saw in the wild crows that lived locally from this point onwards was incredibly interesting and totally surprising. At first, the crows responded in a predictable way to this unpleasant bit of manhandling. Crows that had previously been captured demonstrated typical scolding responses to the male or female researchers wearing the caveman mask, something the researchers referred to simply as 'individual learning'. The crows remembered the encounter, in other words. It had logged in their brains. But then things started to change … The researchers started to notice that other crows began to scold them when they donned the mask. Even if they hadn't themselves been manhandled, they still protested. And then more protested. They too would caw and scold and gather and mob. And then more and more crows would join in. The implications were clear: the crows were passing the information about threats on to one another. The birds were displaying horizontal learning, something previously thought to be mostly a primate thing.

But there was more – individuals also passed the information on to offspring (called 'vertical learning'). News was spreading throughout the local corvid population and it was about a threat: *watch out for the bastard with the mask.* There was a kind of culture, in other words. But still the information about the caveman kept on spreading. Within five years, the 'learning enabled scolding' expressed in those first mishandled crows had spread at least 1.2km from the place of the original mishandling. The crows had watched one another's responses to novel threats. They had learned from one another, just like us. Crows use a simple low-risk mechanism for transmitting information about threats. They don't need knowledge about the severity of the threat; they just learn from one another. They just take it in. To them talk is cheap, in other words. It was a truly fascinating finding, and further evidence of just how wily and impressive corvid cognition really is. In fact, for me it is one of the most intriguing (and almost chilling) discoveries in animal behaviour in recent years: that crows pass on information about threats. They pass on information about the life-threateners; they help one another live by gossiping about the things that might kill them.

Naturally, upon hearing about this research with crows I wondered whether we're the same with spiders. I have no evidence for this, it's just a hunch really. But seriously? I mean, look at our behaviours … We humans have an inherent interest in life-threatening situations. We like telling stories about it. We like passing it on. And it's not just spiders I'm talking about here. Pick up a newspaper: you'll read about disease, war, plane crashes, earthquakes, drownings, murders, suicides, injuries, car crashes and plenty else to do with death. People who say death is taboo are mostly wrong. We *love* talking about death. Or rather, we love talking to one another about novel threats to life. We like sharing it. It brings us together. We learn from it. The study on the American crows really did capture my imagination on the subject. And when it comes to spiders I can't help feeling we are the angry mob of crows soaring around the field, sensitive to the threats expressed by our peers and our parents, but unable to perceive properly the real and present threat to ourselves. Every Facebook share about deadly spiders is a kind of 'CAW'. Every headline in a red-top newspaper a 'whoop' or a 'scold'; a cheap information exchange that is so easy to share it doesn't really matter how true it really is. We are the crows whooping and scolding and circling and John the journalist was my man in the mask, the person who discovered how easy it was to acquire readers and brew up a storm. A cultural explosion. A man who stumbled, for a few months and years, upon a gold mine. A journalistic trope that preys, unlike the spiders, on us and on our instinctual habit for gossip about death. A trope infused with death. A trope that we probably haven't seen the last of – and won't any time soon. Watch out for it. And try if you can not to join in.

CHAPTER NINE

Suicide, Snowy Owls and the Executioner Inside

It was the morning of 8th November 1946. It took place in the middle of the Atlantic on a routine journey from New York to Gibraltar. The crew of the USS *General LeRoy Eltinge* were out on deck, staring perplexed at the pair of objects that had landed on the radar aerial overnight. Big, white, fluffy – they looked like they belonged there; that they do this all the time, locate ships in the ocean to have a bit of a break from all the flying. The captain was thankful that they had chosen a particularly large radar aerial to land on; though these rotund ghostly owls were heavy and imposing, this particular radar aerial was pretty strong – they were causing little disturbance to his delicate instruments. In fact, they may have been there for a while without anyone knowing. Still, they couldn't carry on the whole trip with two snowy owls sitting on the aerial. What if they flew to another aerial and damaged something? thought the captain. What if they broke his ship? No, that wouldn't do. They had to go. The captain ordered the ship's horns to sound. Ahh, brilliant. That worked, he thought. The owls quickly took to the air. They flew around for an hour or so, just long enough for the crew to wonder if they might leave forever. But then they returned. The captain tried again with the whistle. The owls flew around a bit but then returned. And again. And then again. The USS *General LeRoy* was the only object worth landing on for hundreds of miles – the owls weren't going anywhere. The records don't say what happened next. They probably carried on with their voyage, their mysterious cargo – the snowy owls – lost, alone and hungry.

But this wasn't the only boat to have had snowy owls perch upon it that year. Other ships reported similar encounters. Three hundred miles south-east of Newfoundland, for instance, the transport ship *James Parker* was boarded by a lone snowy owl. The owl sat on its foremast contentedly, to the bemusement of the crew. Then there was *Acorn Knot*, heading from Nova Scotia to Reykjavik, 500 miles out to sea. Again, it was the same story: a snowy owl emerged from the mist and perched itself on the boat's mast before later flying off. Overall that year, ornithologists received 24 reports of snowy owls on ships. What in God's name could have caused such a phenomenon? The answer was lemmings.

Lemmings are nature's most famous cyclists: some years they live in peace and harmony with nature in the Arctic, mowing up moss and reproducing contentedly, but then … more and more and more they reproduce … suddenly, months later, it becomes utter chaos. There is no longer enough moss under the snow to go around, so starvation ensues. A frantic search for more moss begins and a mass invasion of new territory occurs – a very famous mass invasion. Many, or most, lemmings leave the area and die in their search, leaving their predators to go hungry. On the whole the foxes become wily, turning to other food sources locally. But the snowy owls depend on the lemmings. The snowy owls fly – some head south through Canada towards the US, even turning up in towns and cities, while many others, it seems, take to the sea. Some of these find boats to rest upon. Many, many more will die, falling namelessly into the waves, starved and alone and forgotten. We may not see another year quite like 1946, when USS *General LeRoy Eltinge* and others noticed the mysterious appearance of snowy owls sitting on the masts, but there have been other invasions since. The moss regrew, after all, and the lemmings repopulated so the owls lived well for a bit. And then they didn't again, so off they fled. But one thing is clear in all of this: though 1946 was a particularly terrible year for lemmings and for snowy owls, neither species was displaying suicidal tendencies.

Where and when lemmings got their mass-suicide tag is largely unknown. Stories of their great plagues appear throughout medieval literature, with many people appearing to believe that somehow these creatures, in the good years, spontaneously generated in the heavens before falling from the sky during storms, and later dying in their thousands. 'The Gods they work in mysterious ways,' these medievals may have argued, but I hope that at least one or two of them may have questioned why an almighty being would display such apparent delight in throwing millions of rodents down at them like they were some sort of multi-ball bonus in a great cosmic pinball machine. The seventeenth-century Danish physician Ole Worm first put right the notion that these creatures were created in the sky from nothing. 'Lemmings are real,' he realised, taking his first step into a brave new world where things could actually be real. 'They must be brought here by the wind!' he declared, though I paraphrase. It took the great Carl Linnaeus (the self-appointed 'prince of botany') to see lemmings put correctly in their place as particularly wondrous and exquisite rodents, and nothing much more than that; natural in their origins and natural in their behaviours, except for the whole suicide thing, which was definitely still a bit otherworldly. This popular notion of them being mysterious suicide artists has largely persisted into recent times.

At this point I have a mental exercise for you. I want you to conjure up in your mind's eye a picture of lemmings hurling themselves off cliffs in their thousands. You might find it particularly easy to imagine, because you've probably seen the footage somewhere. The footage of them leaping – hundreds upon hundreds of them – off a cliff is one that has become etched into the cultural zeitgeist. A generation, including mine, that said: 'Lemmings commit suicide: I saw it on TV.' Of course, you'll probably know now that lemmings don't *actually* commit suicide. Instead you'll probably appreciate that these mass deaths are the result of the mass exoduses (during boom years) toward unpopulated areas after local food exhaustion. During this period, individuals of many lemming species may inevitably find themselves crossing rivers, streams and, occasionally, leaping off boulders and rocks in search of less densely occupied locations. Many die in the struggle. But there in your head, you might still be able to conjure up that image: hundreds upon hundreds of lemmings flinging themselves off a cliff and into the water and dying. I can see it as I write this: close-up shots taken from below of them pouring out and over the cliffs. Falling like grains of sand down the cliff's banks. Overhead shots of them tumbling and spinning through the air, splashing into the waters beneath. Mid-shots of lemmings splashing into the water, consumed in slow motion, down into the icy depths of the Arctic Sea. Still they come. Still they tumble. I can see it so clearly and I suspect you can too.

There's a reason you and I might remember it like this. It's because we're imagining the footage used in Disney's ground-breaking 1958 nature documentary film *White Wilderness*, which I suspect came to be stock footage for a host of subsequent nature films. I rewatched it recently. In its original incarnation, the American narrator talks with a dramatic and tight-lipped intonation like something out of a World War Two propaganda film: 'They've become victims of an obsession – a one-track thought: "*Move on! Move on!*"' he narrates urgently. 'This is the last chance to turn back, yet over they go, casting themselves out bodily into space … and so is acted out the legend of mass suicide …'

What's perhaps most incredible (and which I had never appreciated until now) is that those lemmings were probably pushed over that cliff. Yes, PUSHED. In an exposé that shook the world of natural history programming, a 1982 Canadian TV magazine accused Disney of using a rotating platform to force lemmings over the cliff for them to film. Lemmings, the Canadian TV magazine show argued, were piled onto the rotating platform and forced over and over onto one another. Eventually, too overcrowded on their little platform, they plunged over the edge and into the waters below (which, incidentally, were said to be in Calgary, not in the Arctic). It remains unclear how much Disney knew about the whole thing, but what's interesting to me is how clearly the image remains in our heads. And in the cultural zeitgeist. In reality, nature is rarely as brutal as we sometimes like to imagine. Evolved acts of suicide are present, sure, but not in the way popularly imagined.

Most famous, in the zoological suicide stakes, are the spiders. Small males of many species may approach larger females, plug in a sperm-filled palp (a penis-like organ), and then they'll literally throw themselves at her jaws whilst still plugged in. Sure, he'll die (yes, there is that) but the behaviour (if scientists are reading it right) might well be evolved. It pays out in genes, in other words. After all, they say, on paper a female eating a male will be far too busy eating to allow any other male a chance to approach for a sexual encounter. Genes for this behaviour therefore spread, they say. But is it really that clear? I always feel, told like this, that the female sounds like she's being duped. This is unfair on female spiders, because it's quite likely that the behaviour, if evolved, has in fact co-evolved – it has benefits for both sexes. Her offspring will undoubtedly benefit from all the nutrients that the male has within him; he's a provider of nutrients and she, a provider of eggs. And then there is also the possibility that, at least in some species, this is all just some awful mistake – that the scientists are reading it all wrong. It could be that, for instance, midway through sex the male accidently moves his body in such a way that the female's killing instinct is activated and he is instantly impaled on her fangs and eaten. It really could be that simple, sometimes.

What's really going on with all this killing and eating of mates has been a matter of some debate for spider experts, and it continues to be. But now, thankfully, they have a new group of scientists with whom they can exchange notes: the malacologists. For some molluscs may wield this same trick during sexual encounters. In 2014, for the first time ever, this kind of behaviour was observed in the common reef octopus. In fact, it's now been spotted three times in common reef octopuses. Females have been observed to throttle males with a single tentacle during sex and carry their dead prize off to a cave, one presumes to eat them. It may be that, as with spiders, the motivations of females and males to commit such acts will be difficult to unpick. But the idea of a whole new, relatively distinct class of creature, with impressive cognition (which, in some ways, mirrors our own), undertaking such behaviours is intriguing. Incidentally, this observation may explain why some male octopuses have evolved such a long penis-tentacle. It may possibly be a way to keep females at arm's length – literally. The fact that these female octopuses are reported to constrict their mates in such a slow and sensuous manner is an additional bonus for those warmed by the lurid.

Sex and suicide may sometimes be intimately linked (as we learned in Chapter 7), but not always in the ways popularly imagined. It's certainly fair to say, though, that there aren't many examples of animal suicide where sex isn't somehow part of the plan. Sure, I suppose there are the infertile drones of ants, wasps and bees that occasionally stick up for one another in their attempts to protect the nest. But even this is ultimately about sex: about infertile drones protecting the sexual queen, for instance. Other eusocial insects, including some termite soldiers, go for a similar tactic – a suicide approach called 'tar-babying' (named after the fictional character in the nineteenth-century *Uncle Remus* stories). These soldiers intentionally rupture special glands in their body to produce a sticky secretion that immobilises enemies like ants – they basically blow themselves up. Even they, though, like the ants and wasps and bees, have their eyes on the prize: genes. It seems that if dying helps your chances of sex and replicating your genes, intentional dying – suicide – may evolve.

But it's not as simple as that. Watch some creatures and you will see that sex apparently ceases to occupy their minds. They forego food and walk like zombies up blades of grass, eager to be eaten. Eager to die. They swim round and round in apparent ecstasy near the surface of the water, eager to be consumed. They have a death wish. They are desperate for death. These animals have suicidal tendencies, not to further their own genes but to further the genes of the parasites that have taken up residence in their bodies. They are now being mind-controlled. This is nature at its most awesome.

Toxoplasma is a classic mind-controlling parasite, with an interest in causing suicide. If you're reading this on a crowded bus or a train, then it's likely that the person on your right or your left has *Toxoplasma*. Or you, of course. Toxoplasmosis is a disease which you will probably have heard of if you have cats. Under a microscope, *Toxoplasma* looks a little like a tiny sausage. And it is tiny. This tiny protozoan has a thing for members of the cat family because cats are the only vehicle in which *Toxoplasma* can fully become sexual and have sex. It spends its life moving from host to host, up and down the food chain, awaiting an opportunity to FINALLY end up in a cat's digestive system. In their wildest dreams (if they could dream), all of them hope to one day end up in a cat's intestines. That's where the action happens. It's all they aspire to be: a protozoan that has sex in a cat's intestines. Most don't make it, of course – they lie (as oocysts) in the soil and dirt and water, hoping to contaminate a rat, a mouse or (worse) livestock or (worse still) us. But getting into one of these animals is ok since the *Toxoplasma* can manipulate its odds of getting into a cat *by helping its hosts get eaten by cats*. What *Toxoplasma* does to mice and rats is the stuff of legend. Rodents suffering with toxoplasmosis start to become very unlike rodents. For a start, they begin to show less fear of cats. And they actually start to seek out the smell of cat's urine, drawn to it like the song of the sirens. There are other changes, too, that *Toxoplasma* inflicts on its hosts. Infected rats cover greater distances when they travel. They explore more; their anxiety responses are hacked into by the parasite and they race into uncertain situations like an insecure drunk freshman eager to make new friends.

As I mentioned, it's highly likely that someone you know has toxoplasmosis, or perhaps you yourself – once they're in, they're in. In most healthy people, the protozoan quickly enters a latent phase; it forms tiny cysts in nervous and muscle tissue and sits it out, hoping upon hope that you may be eaten by a lion or a tiger or maybe a clouded leopard or a serval or something, which is possible but distinctly unlikely. For the vast bulk of *Toxoplasma* in human hosts, they've bet on the wrong horse. Realistically, we are dead ends in the great travelling life of this absurd and ridiculous protozoan. Worldwide, perhaps 30 per cent of us carry them. Thankfully, in this latent state, they do little to impact on our lives (though some people report flu-like symptoms at first). If you are an infant or suffering from weakened immunity or pregnant it can be serious, however.

Strangely, there's evidence that *Toxoplasma* can mess with human minds in a similar way to how it acts on rodents. There's evidence that infected people may be less bothered about the smell of cat urine, for instance, and, believe it or not, there's also some evidence that infected human individuals are more outgoing and dress differently from uninfected individuals. No matter how you look at it, *Toxoplasma* is an evolutionary dynamo. No doubt about it. A dynamo. Where other parasites, like fleas and pubic lice, have been persecuted into oblivion by humankind, *Toxoplasma* has ridden the pet industry bubble – and it is a bubble that doesn't look like bursting any time soon. *Toxoplasma* got in on the ground floor, before domestic cats hit big on Earth, and like a Trojan Horse it has spilled its soldiers into our warm and cosy lives. Incredibly – and no one has any idea how this has happened – it is even found in New Zealand's Hector's dolphins. This is a story much more engrossing and real and terrifying than a bunch of lemmings jumping off a cliff in Calgary, at least in my view.

So, back to lemmings. The lemmings really do jump from the cliffs in times of chaos. It really is chaotic behaviour and this is mainly because lemming populations don't behave normally. Unlike in many rodents of Arctic regions, lemming population crashes look jagged on a graph. In lemmings one doesn't see the smooth peaks and troughs seen in most rodents, caused by rising and falling populations of predators. In lemmings it's much more spiky: UP, DOWN, CRASH, BOOM. Jagged, in other words.

The reason for this is because lemmings aren't thought to be limited by predators. They are limited, on the whole, by food. Too many lemmings under the snow nibbling the moss, hidden from predators, means suddenly that numbers can boom. A time of plenty then becomes a time of sudden sparseness. So what else is there to do but run? They run and that's when the chaos ensues. But it isn't suicide; not like it might sound, anyway. Animals don't, on the whole, commit suicide for reasons other than sex. Animals, on the whole, don't kill themselves unless their parasites want them to.

But there is another way to think about suicide. In many ways suicide is, actually, very common in nature. In fact, suicide is incredibly common in nature … it's just that it happens within us. Within our bodies. It happens within us every day. Every hour of every day. Every minute. Every second. Cell suicide. We have had to become masters of cell suicide to live in such complex bodies. Without mastering cell suicide we would see the same chaos that is evident in lemmings: the same overpopulation, the same riotous spreading, and the same illusion of a broken society. We would be multicellular beings out of control.

Caspase enzymes are the key mechanism through which cell suicide occurs. It's easy to imagine these enzymes acting rather like cellular kill-switches. Nick Lane (in the brilliant *Life Ascending*) describes caspases as acting in cascades 'in which one death enzyme activates the next in the cascade, until a whole army of executioners is let loose upon the cell'. Caspases are essentially the mechanism that delivers programmed cell death, a process called apoptosis. Apoptosis is incredibly valuable to life in multicellular bodies. Famously, the cells between our fingers and toes, when in embryonic form, are killed off through apoptosis.

Our bodies are cut to shape by it, essentially, and caspases are the mechanism through which this all happens. Apoptosis accounts for the loss of a tadpole's tail, too. In fact apoptosis is observable a great many times during development. Apoptosis removes the teeth buds from baleen whales, for instance; it also gets rid of the pelvic rudiments of snakes and it almost totally eliminates evidence of the existence of eyes in embryonic moles. But to consider apoptosis as simply a developmental sleight of hand would be to downplay its significance dramatically. For apoptosis is something you are performing right now without knowing it – whilst reading the last three sentences alone you've destroyed about 10 million cells through it. Oh, and there go another two million. Well done. Every second your various cells are being snipped up into fragments that are digestible to phagocytic cells with minimal fuss and damage to neighbouring cells. That's apoptosis.

Apoptosis is a must-have tool for multicellular life. Without it there would be no us, no fish, no bats, no frogs, no lemmings. And why? The answer is simple: apoptosis stops cells going rogue, dividing rapidly and causing, in the worse cases, cancer. It finds and murders the single-celled transgressors trying to break free from our multicellular approach to life. It kills them. And caspases are its weapons of choice in this battle. They really are the executioners. Cell death is an everyday thing, something that we are all masters of. You owe your life to it. Almost every cell you had in childhood has gone, mostly through apoptosis; nearly all of your body's cells have been killed off and replaced. Even when old, we live our lives in bodies made, mostly, of new cells. Our totality is within generation upon generation of cellular blooms that form in the wake of cells that have died. Blooms controlled fastidiously by death. By caspases.

Where and when multicellular life evolved the ability to wield caspases to enable cell suicide is anyone's guess, but it must have happened very early in evolutionary history because of one curious observation: caspase enzymes are found in us and they are also found in cyanobacteria. Actually, this makes total sense. Caspases aren't used by our cells at all, but actually by our mitochondria – the ancient symbionts that live within our cells, with a deep history as once free-living single-celled organisms (prokaryotes) tamed and given pasture within our eukaryotic, multicellular selves. The writing is on the wall: our success as multicellular life forms isn't because of us, but because of our mitochondrial hitch-hikers, possessing the tools to cut back the cells that behave and replicate too wildly. In other words, we have mitochondria to thank for each and every day that we are alive. But there is more to the story than this. For instance, how and why did mitochondria (or their ancient free-living relatives) evolve the ability to commit suicide through caspases? Why would natural selection select for such an ability? It makes little sense in Darwinian terms, for single-celled organisms to go around killing themselves, after all. Thankfully there is one organism that may offer us clues. It is called *Trichodesmium*, and in many ways it might offer us an insight into why such ancient organisms required the tools for suicide.

Great blooms of *Trichodesmium*, a cyanobacterium, are occasionally known to cover hundreds of square miles of open ocean, making the sea blood red in places (in fact the name 'Red Sea' probably comes from *Trichodesmium*). These blooms (each made up of floating single photosynthetic cells) can last for weeks and perhaps months, thriving on minerals carried from rivers or via upwellings from the deep sea. But then they die. The caspases are activated within their cells. All of the *Trichodesmium*, almost overnight, cease to be. So why? Why do they commit suicide? It's still largely a mystery, but the answer is probably something to do with viruses. Today and perhaps for billions of years, these enormous algal blooms are preyed upon by an army of viral parasites, each eager to break into cyanobacterial cells and reprogram them into their own unique brands of virus factory. There are rather a lot of these viruses out there, as it happens. In fact, a litre of surface seawater contains about 10 billion virus particles. So why do the cyanobacteria kill themselves? One guess is that when the nutrients start to dwindle even slightly, it may pay more in surviving genes for individual cyanobacteria to pull back from the advancing army of viral parasites, hobbling the virus's frantic evolution into ever-new forms of perniciousness. The viruses fall back whilst some of the cyanobacteria sit it out on the floor as hard cysts and revivify themselves later, possibly when the viruses have washed away or become denatured somehow. In theory, these new individual *Trichodesmium* get a head start before the viruses re-evolve once more to cause havoc. And they will gather again in huge numbers, of course, as they have probably done for billions of year.

To *Trichodesmium*, a healthy sprinkling of death puts the viruses back in their place, allowing the cyanobacteria a head start when the good times return. As with the lemmings there will be peaks and troughs, sure, but a bit of die-back can sort out the predators and allow the nutrients to build up once more. It can be healthy to kill yourself, in other words – provided some of your relatives manage to encyst on the seafloor and live and blossom again. And so it could be that the ancient free-living ancestors of mitochondria that gave rise to those that inhabit our cells went through a similar viral assault before nudging their way – quite literally – into our lives. It's going to be difficult to prove this, of course – but it's a tempting hypothesis. I for one love the idea that we multicellular animals can persist in multicellular form because of an ancient war fought in the sea between our mitochondria and viruses.

We owe the mitochondria for our weaponry, and the mitochondria might owe the viruses for theirs. Of course, we may never know. What's clear is that the evolution of cell suicide had a dramatic impact on life on Earth. We couldn't live without it. It allowed us to build order in otherwise selfish aggregations of single cells through wave after wave of execution. Through such cell suicide, powered by mitochondria, multicellular life suddenly gained a hand in survival. Those early twigs of life in which mitochondria featured grew fast and gained strength and would come to form three great bows of multicellular life that exist today: the plant kingdom, the fungi and, of course, animals – each prolific wielders of caspases, courtesy of their mitochondria. Each prolific wielders of death. Organelles that work together to pull back from chaos; that pull back from the spiralling madness of disordered cells that appear like a hundred thousand lemmings fleeing outwards, leaving behind them emptiness, starvation, bodily death; no order, no aggregation. A true and utter chaos. Mitochondria deserve more of our time. They deserve more of our thanks. Mitochondria provide our energy, they provide the weapons through which we can keep our multicellular bodies from becoming unicellular, and, possibly, they possess properties to keep free radical damage at bay. Through this, possibly, they may contain the mother lode: immortality … or something a little like it. Screw lemmings – if anything deserves the Disney treatment, it's them.

CHAPTER TEN

This is Not a Sheep

From over the hills to the left they had loomed. At least a hundred at first. Then two hundred. Looking up higher into the atmosphere I had seen more. Five hundred. Still just dots, really. Five hundred dots. They formed an enormous spinning column, these dots, like distant wreckage swirling within a tornado. Debris with menace. There was a faint mewing from the sky, so light as if to be almost imagined. My family, part of the audience, were within a crowd of perhaps three hundred people that had gathered specifically to see this, arguably one of the western world's most impressive gatherings of scavengers. There was a small lake in the middle of the bowl in which we sat; our conversations and nervous laughter, as the dots came nearer, echoed over the water, our noises held in by rows of dense pine trees that overlooked us on all sides. An echo-chamber. After a few minutes they came nearer still. We sat underneath them, as if in stands, like an expectant crowd. It really was like a stadium, and we were the expectant hordes, baying for blood. We looked up. A few minutes later it looked like there were thousands of them. They were getting closer. They were red kites.

In recent years seeing the red kites has become THE THING TO DO in Wales. The 'feeding station' (as such sites are called) in which we were sat at Bwlch Nant yr Arian was designed to give small numbers of the threatened red kites 'a helping hand' in the late 1990s. Now there seem to be thousands. Red kites had faced incredibly serious declines across Britain, and this was deemed a nice way to help their resurgence at one of their strongholds. And it has helped them locally. Each day (including on Christmas Day) hundreds of kites from miles around visit this site to congregate and feast upon offcuts presented by a local butcher. It works like this: a nature reserve manager carts the butcher's meat (in a wheelbarrow) across the site to a little well-maintained mound in the middle so the audience can clearly watch the birds coming in. The scraps are thrown across the mound, and the birds take it in turns to get their fill. My youngest had not yet been born at the time we visited, but Lettie (then a baby) and my wife Emma were there. We had perched ourselves on soft mounds of moss on the hillside as the soaring birds of prey made their approach. There was a buzz about the place, like a music festival about to start; people sat on picnic blankets and nibbled homemade sandwiches, and many had their cameras ready and pointing at the mound on which the offal was to be presented. It dawned on me that many of the nice shots I've seen of red kites in magazines were probably taken here, where it's incredibly easy to take exceptional photos of red kites since there would be about 1,000 opportunities to see them up close looking charismatic and enigmatic and swoopy. Here, as I would come to see, taking good shots of birds of prey would be like shooting fish in a barrel.

Through my binoculars I could see, up the hill by the car park, the offal being prepared near the visitor centre. Great bags of it were being tipped into the wheelbarrow as the red kites circled ever closer. I looked toward the sky again. Their charismatic forked tails were now clearly visible, and they lazily flapped on lanky wings, biding their time. My appetite was thoroughly whetted, not least from the buzz on TripAdvisor. '*I was not disappointed! It was a truly mesmerising experience!*' said one of almost 200 five-star reviews we'd looked at the night before our visit. Words like '*Breathtaking!*' and '*Awesome!*' and '*Clean toilets!*' littered the comments pages. '*Even our dog was transfixed …*' said one reviewer. (Imagine being a small dog surrounded by 1,000 birds of prey. Would you be transfixed? Yes you would.)

Then there was activity. When the wheelbarrow of offal was being wheeled down, there had been a sudden change in atmosphere; these enormous birds came suddenly lower, right above our heads, weaving within trees and ghosting over the edge of the grassy arena like great pterosaurs from the Mesozoic. Their shadows flashed over the crowd like angry spirits. It was wonderful. A real spectacle was beginning. But then things took a personal turn for the worse. Emma gave me a little nudge. She nodded at the baby, who had chosen this exact moment to soil herself in a particularly violent way. The smell was not very good. This was a problem for us: the toilets were about a half a mile away and whoever went would miss the whole damn thing. But there was another option: the baby could be changed high up on the hillside behind one of the pine trees, away from the crowd and where no one could see. I looked at Emma pleadingly. Could she go? She had pointed at her ridiculous sandals, completely inappropriate for climbing up the wooded hillside behind us. 'That's fine. I'll go then,' I'd mouthed so that the people around us couldn't hear, doing a terrible job at hiding how annoying this all was. 'That's fine,' I hissed. I gathered up the nappies, the wipes and the soiled baby and started walking up the hillside into the dark forest at the exact moment that the roadie with the wheelbarrow of offal came down the long path toward the sacrificial mound. Pandora's box was about to be opened and I would miss the whole thing … unless I hurried.

I scurried up the hill with her in my arms, as the devil birds made their approach. As I looked for footholds on the steep banks beneath the trees I could hear them now. Calling. Hundreds of them. Mewing from all around with a penetrating excitement. I guess it was hunger, but it sounded almost like song. Out of breath and sweating I found a suitable moss-lined cleft behind a tree, high up and far away from the crowd. I began changing the baby. I could hear excitement from the crowd below; I guessed that offal was now being unloaded, thrown in great chunks onto the mound that was just out of view behind trees. The red kites had started making their dive-bombs. It was definitely beginning. I tried to focus on the job in hand …

There is a reason I am telling you a story that involves my daughter's faeces. It's integral to the story. The faeces is what made me come up here, away from everyone else, remember? And from up here the scene was magical because, at the top of the steep bank looking down, I was at the same altitude as the birds. I could see what they saw. My daughter and I were in the middle of the pack; red kites above us, red kites below us. We were among them. The kites came screeching between the trees, sometimes almost at arm's length from us. They weaved between the branches, their wide lanky wings locked open, arched and rigid like those of a starfighter. It was like being in the middle of a beautiful war. And I could see the crowd from up here so clearly. The picnickers, the photographers, the nervous dogs. They were a proper crowd. An audience. We stood watching them. The crowd began whooping with joy, laughing, smiling. Drinking fizzy drinks. Opening bags of crisps. This was entertainment to them; by paying a small car-park fee they had become paying customers, and this was their reward. No wonder this place had got such brilliant reviews, I thought. The visitors here could see and admire something from the distant past. Hundreds of scavengers, scavenging. Doing what they used to do in Britain hundreds of years ago. Scavenging – a behaviour we rarely get to see in Britain any more.

And it started to hit me. The whole thing … the crowd, the birds, the wheelbarrow … it was a scene that didn't quite belong in the modern world. It dawned on me that this was a kind of theme park. A novelty experience. A day out. And had I been down there, amongst them all, I would have failed to notice any of the cultural significance of it. I would have been caught up in observing these wonderful birds for their grace and their power and their beauty, oblivious to the human perspective. I would have missed seeing the humans and the kites juxtaposed against one another. I realised at that moment that what I was watching spoke volumes about a world we once had and a world we've mostly lost. For aerial scavengers are amongst the fastest disappearing creatures on Earth, and theirs is a story that deserves a chapter of their own in this book. And so here is my best attempt at that story.

Red kites are one of those literary birds from previous centuries that we almost forgot about and that we almost lost forever. Long, sword-like wings and a deep V-shaped tail; they fly as if in a perpetual state of indifference. They are graceful birds. Although not total scavengers (they occasionally take small mammals and reptiles and even earthworms), they are an animal that was once closely associated with death across much of western Europe, particularly in medieval times when they became creatures equally at home on sewage tips or on battlefields. In Shakespeare's *Coriolanus* London is referred to as 'the city of crows and kites', a reference to the commonness of such a large and prominent bird. In *The Winter's Tale* Autolycus says: 'When the kite builds, look to your lesser linen' – a reference to the apparent phenomenon of kites stealing clothes from washing lines to furnish their nests (indeed, so common were these creatures centuries ago that they were known also to 'snatch bread from children, fish from women and handkerchiefs from hedges', according to the English naturalist William Turner, writing in 1544). It's probably not fair to say that the birds were revered, exactly, but red kites served a useful purpose in those times as hook-billed street-cleaners (unwanted scraps? Let the theropod dinosaur with a six-foot wingspan come to your door and take them away!).

Things changed in a big way for the kites, though. In the mid-fifteenth century, as the agricultural industry blossomed in Britain, the lazy flight of the red kite took on a more sinister note. In the eyes of the elite these birds ceased to be collaborators in nature's great story. Suddenly they were competitors. It was deemed that they wanted what the rich had: livestock. Their hooked beak and strong claws were all the evidence needed – they were killers. No doubt remained. Though they were previously given a degree of legal protection (for, God bless, services to public hygiene), the knives were sharpening. In 1566 the law changed and the tide turned for red kites in Britain. Rather than celebrate such creatures, almost overnight the kites became legal vermin. As with jays, ravens and (strangely) woodpeckers, the killing of kites by rural folk was now to be positively encouraged. It was to become a kind of sport, a bit of fun. No surprise, then, that at the end of the nineteenth century there were barely any red kites left at all in Britain. Many of the kites was killed brutally. And for those that remained, things got even worse. In the twentieth century legislation to improve biosecurity was introduced: many farmers began to remove dead livestock from fields to reduce the spread of livestock diseases, and so one of the kites' main sources of food, carrion, was being taken away.

It's a wonder that any red kites survived at all, but some kites got lucky – a relatively untouched UK stronghold remained and with it so did hope. This small population was in central Wales. It had miraculously clung on, helped by volunteers who stubbornly guarded nests from egg-collectors who, for a short while, threatened their very existence. This is where their story picks up. Over the decades that followed this population stabilised; in fact, it stabilised so much that conservationists realised that this central Wales population could be used as a breeding ground from which other sites around Britain could be repopulated. So, in 1989, conservationists began on an ambitious project to do just this. By translocating birds from Wales (as well as Sweden and Spain), new populations have now been established in England in the Chilterns, Northamptonshire, Yorkshire and Cumbria. Northern Ireland is another place where reintroductions have had encouraging results. As of this moment there are 16,000 pairs of red kites in Britain and, for the first time in centuries, red kites are now occasionally spotted over the City of London, just like in olden times. They're coming back, in other words. Reclaiming their niche: death. Though some are still shot or poisoned (illegally) by landowners and gamekeepers, the red kite has become a global success story for people, like you and me, who like the thought of big creatures living wild. It's a story with which all conservationists enjoy regaling their children, whenever red kites are spotted circling high above on long car journeys: 'When *I* was a kid there were almost none left …' we say with great pride. We don't have many conservation success stories in Britain; this is one we like to celebrate. This is one of Britain's biggest remaining aerial scavengers. And every continent has its own. For now, anyway …

Among these continental cliques, of course, the vultures are the ones we recognise best. Books and TV programmes tell us that these, truly, are creatures that have evolved an almost perfect design to feast upon the bodies of the recently deceased. If red kites are refuse-collectors, then true vultures are more like mobile industrial chippers; homing in on freshly decaying matter and turning it into mush in minutes. Broadly speaking, vultures aren't a single family; rather, they are two, spread across every continent except Australia and Antarctica. One, which biologists divide into two closely related subfamilies, lives in the Old World. The other, the New World vultures, have sprouted from a different part of the avian family tree entirely, but have evolved convergently many of the same features as the Old World vultures.

Among the most celebrated of the vultures' adaptations for life on the corpse are their bald heads. Famously we're told that the vulture's head is largely devoid of feathers because, well, sticking your head into a blood-spattered ribcage is not conducive to maintaining good head-feather condition. This is one of many untruths about vultures. The reason for their baldness may actually be far more prosaic. Some scientists suggest that it may be thermoregulatory – having a semi-bald head may be a simple mechanism through which to reduce the likelihood of your face (or brain) cooking in the hot sun. And, fair enough, it's true: many vultures do live in warm climates.

Vultures have other adaptations for life on the bones. Their stomach acid is exceptionally corrosive, for example. Helpful for digesting smaller bones, certainly, but probably also useful for destroying micro-organisms that associate with putrid carcasses, like *Clostridium botulinum* (which causes botulism), hog cholera and anthrax bacteria. In 2014, an analysis of black and turkey vulture digestive systems revealed that their gut micro-organisms differ markedly from our own. Where ours house a veritable mishmash of micro-organisms, vulture guts are dominated by just two types: *Clostridium* (which produces botulism toxins) and the dreaded *Fusobacterium* (implicated in a number of serious blood infections). These produce poisons. Poisons that the vultures appear totally fine with. What are these poisons doing there, you might ask? No one is particularly sure. It could be that the vultures have evolved an immunity to them, or that possessing stomach acids 10 to 100 times more corrosive than our own means that these poisonous micro-organisms can't get up to much. Scientists are still looking into it.

There are tales of New World vultures possessing another neat trick to do with death: the ability to projectile vomit their stomach acid when threatened by ground predators, which sounds rather impressive if only in a frenzied *Aliens* sort of way. As with the bald head, the truth is probably a little less exotic, but it may have evolutionary significance nonetheless. After all, heavy vultures take longer to take off, so it might be that vomiting is a bit like throwing out the ballast before take-off. Scientists aren't too sure about this one either. And there are other evolutionary quirks that vultures possess. One is that they frequently urinate all over their legs; a behaviour thought to be adaptive since nothing removes potentially fatal corpse residue quite like uric acid (murderers take note).

Oh, and urine is cooling (apparently) too (murderers: your call). But it is their adaptations for locating dead megafauna that have caused most stir and argument among zoologists. After all, how exactly do vultures get to a dead body so quickly? And from how far might such vultures travel? And what senses do they use to locate these dead bodies? These were once big questions for Victorian naturalists. So heated did the discussion become that it was almost inevitable that two opposing groups would form: one arguing that vultures used smell (this group became the 'Nosarians') and another group arguing that they used sight to locate dead and dying prey (the 'Anti-nosarians'). How they resolved this argument is worth explaining, not least because it outlines the wonderful manner in which great scientific battles can be won and lost through simple experiments (I retell it here recounted from Benjamin Joel Wilkinson's excellent – and freely downloadable – book *Carrion Dreams 2.0: A chronicle of the human–vulture relationship*).

Leading the charge for the Nosarians was the British naturalist and explorer Mr Charles Waterton. Waterton believed that vultures 'snuff the smell / of mortal change on Earth … / Sagacious of the quarry from afar …' But the New World had its own talismanic truth-seeker when it came to vultures. The chief Anti-nosarian was no less than the noted ornithologist John James Audubon himself. The reason for his conviction about this was simple. Audubon had previously observed that a vulture had once approached when he had covered a bale of hay in deer skin. The vulture had approached the strange-looking hay bale and pulled at the 'flesh' ('much fodder and hay was pulled out'), apparently confused that the corpse had turned out to be a ruse. There were no smells of death yet still a vulture had appeared. Audubon had thought about this. Audubon built his hypothesis: he was to become a powerful Anti-nosarian. Vultures located corpses by sight, he proposed.

Ermine caterpillars wield silk like a shield. This allows the caterpillars beneath to continue 'farming' the tree safely protected.

Above: Western scrub jays respond in predictable ways to the sight of their dead. (a) mock-feathers; (b) a dead bird; (c) stuffed owl; (d) stuffed bird.

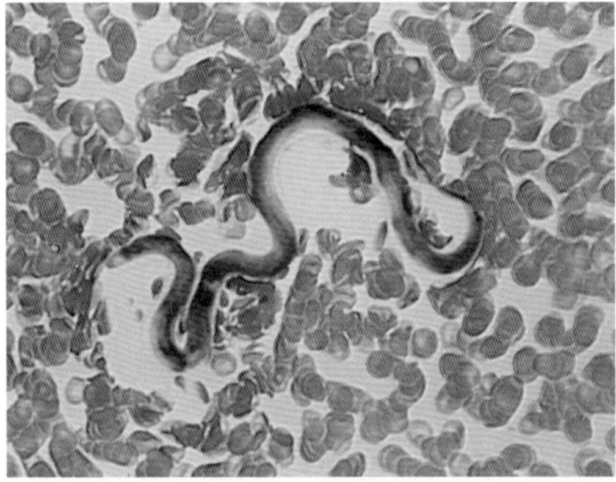

Left: Nature at its most brutal: Loa Loa, a tiny worm known to infect eyeballs.

Left: Pacific salmon are perhaps the most arresting example of semelparity. Hundreds of thousands die each year within days of spawning.

Above: Elephants: our most celebrated animal mourner. But what, really, do they know of death?

Below: A yellow-footed antechinus mouse. All twelve species in the marsupial genus *Antechinus* practice semelparity. Males invest nothing in body maintenance and everything in sperm.

Above: Cookie, a male Major Mitchell's cockatoo residing at Brookfield Zoo, Chicago, is the oldest bird in captivity. Cookie hatched from his egg in 1933.

Below: Insect nutrients re-packaged within bat guano for human use.

Right: 150 million-year-old fossilised faeces, probably from an ichthyosaur.

Below: A riot of rove beetles. Diverse, colourful, oft-overlooked.

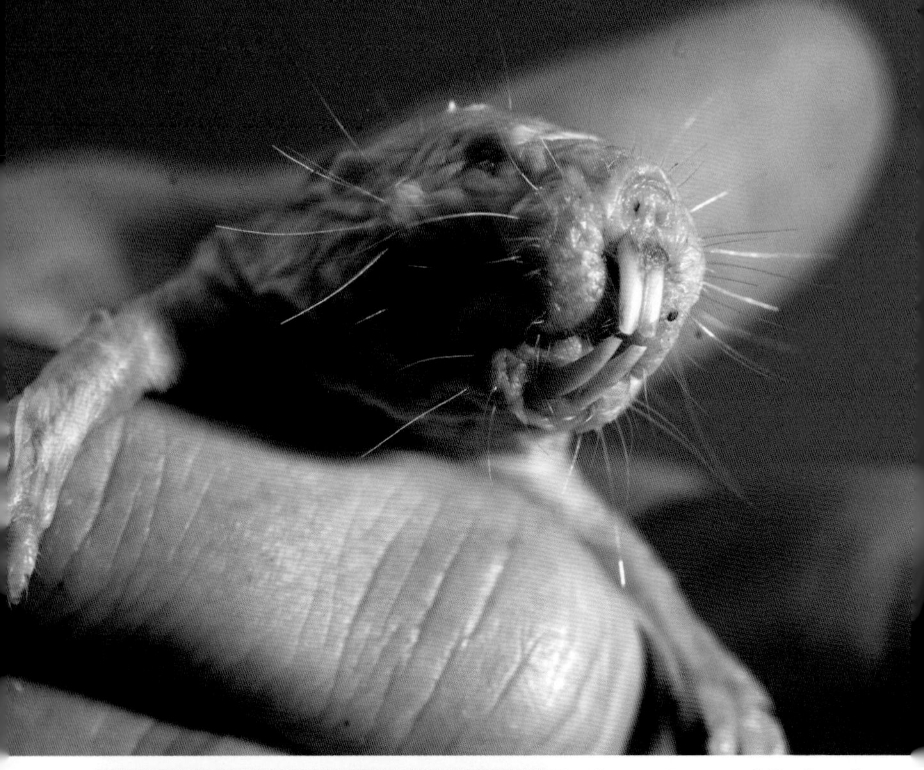

Above: Even riddled with free-radical damage, naked mole rats can live 30 years.

Left: Caterpillars aren't agents of death ... they harbour new life. Within the cocoons spun on this hornworm, adult parasitic wasps are readying.

Left: Common toads: thousands are squashed by cars on their long migrations to breeding ponds. Conservationists generally meet such losses with sadness. But were the lives of these toads wasted? Ask the scavengers, who make such losses their personal gain.

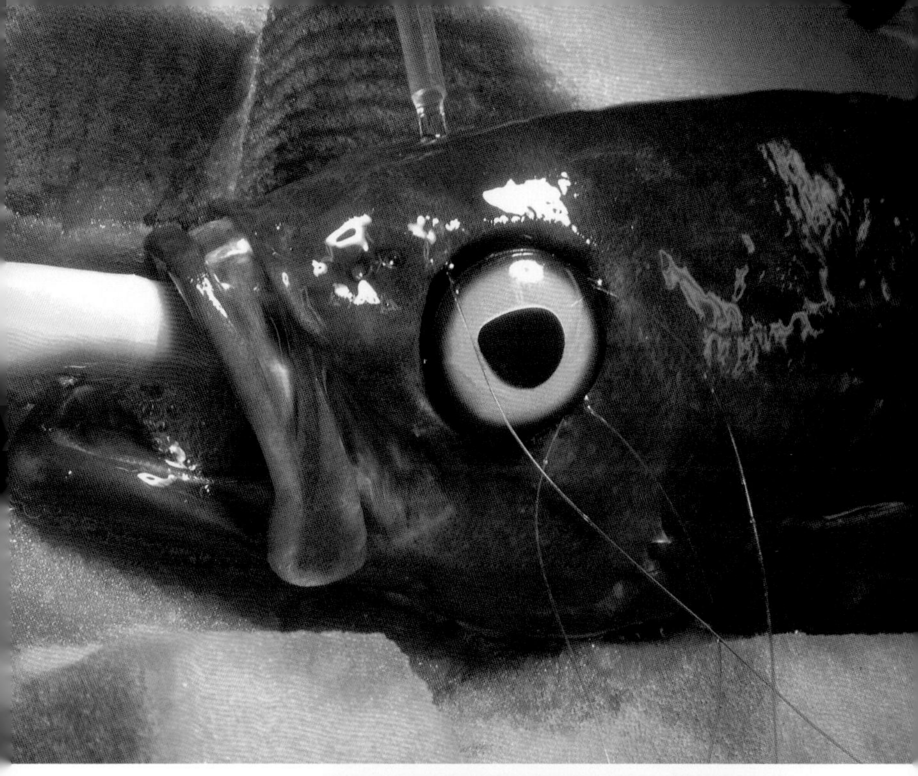

Above: The googly-eyed rockfish ... post-cataract. Fish age like we do.

Right: Dead frogs, post-copulation: victims of evolutionary and ecological circumstance.

Right: *Hydra* exist in most freshwaters across the world. Their stem cells show impressive capabilities for regeneration. Could we one day make use of their genetic trickery to influence our own ageing process?

Above: Red kites, like vultures, offer a valuable death-removal service. They have been clearing up after us for thousands of years.

Below: BEHOLD, THE OLDEST ANIMAL IN THE WORLD. Ming (inset) died at the ripe old age of 507 years. In the background, an Icelandic sea-floor packed with quahog shells.

Genus: Arctica islandica
Ref #: 061294
Locality: Iceland
Station: B05 AD03
Dredge/Tow #:
Ave. Latitude: 66° 31,59N
Ave. Longitude: 18° 11,74W
Water depth: 83-81 m
Collector: Scourse
Cruise: Bjarni Saemundsson, Iceland B05
Collection Date:
Method: Arctica dredge
Live(Y) or dead(A/R/L): YA
Length: 86.9 mm
Height: 72.5 mm
Max Height: 82.1 mm
Width: 48 mm
Weight of shell valve: 52.08g
Periostracum: 1
Ligament: 4Shell margin: 4
Bioerosion: 4
Nacre: 1
Sex:: Spent?
Flesh wet weight: 40.28 g
Notes:

Audubon and Waterton met at a formal talk in London in December 1826, and it was from this meeting that the energetic intercontinental debate about vultures raged. From this point forth, the Nosarians and Anti-nosarians each searched for supporting evidence for their own hypothesis whilst searching for evidence to refute the claims of the other. It wasn't pretty. At one point Waterton said that Audubon's ideas were so half-cocked that 'he ought to be whipped'. Insults aside, in the end it was the superior methodological approach of the Anti-nosarians that set the stage for the winner. Buoyed by Audubon's initial observation of vultures investigating bales of hay wrapped in deerskin, Charleston's Reverend John Bachman took Audubon's methodological baton and ran with it, employing the academic services of members of Charleston's prestigious Philosophical Society to help him out.

Together Bachman and his team of philosophers thought about how they could uncover experimentally the truth behind whether vultures (locally, black vultures and turkey vultures) found prey using their eyes or their noses. They had a great idea. First they got their hands on a dead vulture, covered it in rice chaff, and watched what happened. As Audubon predicted, no vultures visited, even though the corpse stank (almost literally) to high heaven. Then they tried something else. They placed some rotting meat under a platform, so it was invisible from above, yet through which air could pass. The smell 'went far and wide' but still, after 25 days, not a single vulture visited. Only dogs. Lots of dogs came. It couldn't be smell that the vultures were attracted to, they thought. They were getting closer.

And then came Bachman's truest moment of creative genius. He commissioned a local artist to paint a picture – a vision he deemed would be sure to attract vultures from the sky: an enormous painting of a plump sheep with its body alluringly eviscerated. Bachman and his groupies took the painting to the meadow, placed it on the ground and watched what happened. And something did happen. For perhaps the first time in the history of the universe a bird came down from the sky to try to eat a painting. Then more vultures turned up to try to eat the painting. Then others. Then more. The vultures 'commenced tugging' at the great work of art. Then they 'seemed much disappointed and surprised' to find that they couldn't eat a painting, and the little rabble of philosophers watching in the nearby undergrowth all agreed that it had 'proved very amusing'. The test was repeated more than 50 times. They tricked the vultures again and again. Even when offal was placed near the painting the vultures failed to locate it, heading straight to the artwork still. It was a momentous bit of methodological mastery (who says art and science don't mix?).

In 1834, Bachman and his team wrote up their findings and sent them east toward Europe. The Americans, led by Bachman and Audubon, had won the debate. The Anti-nosarians had nosed it. The British, led by Waterton, had lost. They had lost … mostly. I say 'mostly' because, of course, things aren't always that straightforward. Further research (without artwork) showed that black vultures do appear to depend almost solely on sight, but that turkey vultures, on the whole, don't. They have a capacity for smell, too, which they do appear to use while seeking items to scavenge. Much to the disgust of the British Victorians, other raptors in the Accipitridae (the taxonomic family in which both Old World vultures and red kites sit), like most vultures, also have a poor sense of smell; they use sight to home in on dead animals.

This is where the similarities end between vultures and Europe's kites. Currently the plight of Old World vultures and red kites couldn't be further apart, in fact. While red kite populations continue to do well (in the UK at least), Old World vultures are declining at a rate faster even than the dodo once faced. Almost overnight (in geological timescales) numbers of these vultures have plummeted.

Literally and figuratively plummeted. Currently 75 per cent of Old World vultures are globally threatened with extinction or are officially Near Threatened, according to the IUCN Red List of Threatened Species. The vulture species worst hit has been Asia's white-rumped vulture: 99.9 per cent of them have gone, according to BirdLife International (who monitor populations and have campaigned fiercely to see an end to such declines). The cause of these vulture declines? It's called diclofenac, a veterinary anti-inflammatory commercially given to livestock. When vultures feed on dead livestock they then die from kidney failure because of an accumulation of the drug in their body. The drug has wiped out 99 per cent of vultures in Pakistan, India and Nepal, yet despite this horrific (and well-documented) decline in Asia, in 2014 diclofenac became available in Spain and Italy. Both European countries are vulture strongholds. Who knows how their vultures will fare.

Diclofenac is the most concerning of a number of threats to modern-day vultures that include persecution (to supply the trade for traditional medicine), habitat loss, the threat of power lines and deliberate poisoning. And it's getting worse. In June 2015, new studies suggested that Africa's vultures are going the same way as Asia's, having declined at rates of between 70 per cent and 97 per cent in just three generations. This makes me incredibly sad. It makes me sad because vultures are fundamental parts of Old World food chains. They are like alchemists, making nutrient-rich fertilising paste from the bodies of dead animals, removing corpses and eliminating their capacity to spread disease. And they are beautiful. They are big and beautiful and I am in love with them, just like I am with the kites. How will they fare? How many vultures will die in the Old World? It's hard to say. We will wait and see, but there is hope that one day they may, like the red kites, have a stuttering resurgence. Vultures have place and purpose. They have poise and value. They are every bit as special as red kites or their New World cousins, which are faring (in most parts of the Americas) better.

Even though it was nothing much to do with me, I am immensely proud of the resurgence seen in red kites. Proud that I sometimes see them over our house. Proud that they exist still. They are evidence that we can change and influence nature in a nice way. A good way. Evidence that we can influence the lives of creatures many (or most) had written off. Evidence that life, given a chance, heals itself pretty well. But the Victorian elements remain. The distrust, for many, of scavengers is still there. In Britain, as I write these words, it seems to me that this uneasy relationship between some landowners and kites is re-emerging. I fear the gentry may view them as threats to the game-shooting industry. Far right-wing parties refuse to believe that they can be truly trusted, having been reintroduced partly because of a legal bill dished out from a bureaucratic EU they no longer want Britain to be a part of ('These birds, they come here from Europe to steal our lambs, blah blah blah …'). I can offer few insights here, but I feel sorry for them nonetheless. They are treated unfairly in much the same way that magpies and crows and ravens are. They, like many of the corvids, are tarnished because their niche involves the dead. It's really not their fault, surely? After all, we cleared the wild animals with which they coexisted for thousands of years, swapped them for livestock and then actively removed death from the system; pulling the rug out from under them. We starved scavengers out, mostly. The kites and magpies and crows that remain are like the unemployed; we are too quick to call them scroungers. We mistrust and label them in the same way.

And that's what I realised most when I saw those red kites at Bwlch Nant yr Arian. As I watched from the hillside after changing that nappy, looking down upon the whole scene, I could see it all. Those kites, hundreds of them. Spiralling down. Forming an orderly queue as we watched and laughed and mocked and ate our prawn cocktail sandwiches below. It was a food bank for scavengers. They were spoiled birds; spoiled versions of the important life their species had once held on Earth. They were once birds with real purpose and poise; a purpose we took away from them. Britain may never support as many red kites as in previous centuries – there's just not enough death out there. So what is there for us to do? For me, I now celebrate each and every one I get the chance to see. And I pray for a tiny bit more death, not life, so that my children will see them too.

Looking out across the bowl, standing amongst the kites with my nappy bag in one hand and baby in the other … it was the first time that I suddenly had misgivings about humans and our perceptions of death in nature; our role as self-appointed enforcers of life and death. I felt it again with the ermine caterpillars. Then I felt it again with the false widow spiders. A deep uncertainty was beginning to run through me about humans and our perceptions of death. Do we fear such creatures because they remind us of ourselves and our own mortality? Are we scared of them because we're scared of death? Or do we hate them because, try as we might, we can't accept the fact that we are painted from the same watercolours; that life and death are equal parts of who we are and how the world works? Is there a part of society that feels differently? I wondered. Were there people out there who love and accept the bits of nature that remind us of death? And there were. There really were.

CHAPTER ELEVEN

The Grotto Salamander and the Guano

The penguin that washed up in the river on the way to Azerbaijan didn't have a name. No one knew much about it really. It was just bobbing around in the water and then fished out by officials. It was an African penguin, native to the southern coast of the African continent. A travelling penguin, essentially. Genuinely. It had been born in Torquay, England, and then, apparently, it had been given to Tbilisi Zoo in Georgia. It had been washed out of the zoo during a heavy flood and had swum 200 miles downstream and been found in the river on its way to Asia via Azerbaijan. What was an African penguin doing in Europe, you might ask? And for what reason was it so very far from home? It is a story, like so many, millions of years in the making. And it involves faeces.

The romantic retelling of this particular story starts in a previous century in a place not far from Torquay: Lyme Regis, a celebrated part of Britain's Jurassic Coast made famous by Mary Anning, the nineteenth-century palaeontologist. The cliffs above Lyme Regis are famous not only for the ichthyosaurs and plesiosaurs that she discovered, but also for the 'bezoar stones' that fell from them. Spiralled and grooved and often stained with black splodges, these unusual stones inevitably became part of the seaside trade. Jewellers made brooches, necklaces and earrings from them and, again inevitably, medicine men bestowed upon them special powers which could be passed on to the wearer.

Unbeknownst to almost everyone at the time, the women were actually being encouraged to wear little fossilised reptile shits. It was true: a market in reptile shit had, unwittingly, begun. Anning was one of the first to suspect that everyone was walking around wearing fossils of reptile shit because, among other things, she found these mysterious bezoar stones in the lower cavities of ichthyosaur fossil skeletons, being readied (once upon a time) for a bowel movement that was never to be. She knew. Others didn't. William Buckland (the Dean of Westminster who went on to be the first to describe the first fossil dinosaur, *Megalosaurus*) agreed with Anning about this. He undertook experiments on a range of bezoars and discovered within them fossils of fish scales and what he considered (rightly, it turned out) to be ink stains from the sepia-bags of cephalopods (belemnites and cuttlefish, mainly).

Needless to say, soon after the news about bezoars spread through Lyme Regis the bottom (sorry) quickly fell out of the bezoar-jewellery trade. But a trade in something else was to bloom. A trade much larger and more profitable, which went on to feed the world for a while. Buckland, still intrigued about the bezoars, introduced his friend, the noted chemist Lyon Playfair, to the wonder of bezoars, or coprolites as they had been renamed. Together they collected samples, which Playfair pulverised in his lab. As Buckland had predicted, Playfair found the bezoars incredibly rich in phosphates of lime, a key ingredient required to refertilise lost soils. Together these men dreamed big, imagining an industry that could restore the face of the Earth for crops using the fossil faeces of creatures that lived millions of years previously and that no longer had a need for their faeces at all because they were all well and truly dead.

They invited to Lyme Regis a distinguished visitor named Baron von Liebig, a renowned German chemist. They pitched their idea to him; von Liebig liked their idea. In fact he soon published a treatise on the subject (with their support) and went on to build a global industry shipping super-phosphates, not from dinosaurs or marine reptiles, but from the animals that survived them: birds. Their guano became highly prized from this point forth and, in the decades that followed, many millions of square miles of the developed world were made green through the trade. And it all began with innocent earrings and brooches of fossil reptile faeces.

This is the romantic version of events, of course. In truth many early geologists and explorers (including Alexander von Humboldt, whose writings inspired the young Darwin) had noticed the fertilising potential of droppings. Others had reached similar conclusions to von Liebig, far earlier; that guano was highly effective as a fertiliser given that it contained nitrogen, phosphates and potassium, which are all pivotal to plant growth. Still, regardless of how it started, in the nineteenth century the remote islands where seabirds gather in their tens of thousands suddenly had a new value.

A global race for their excrement began. It made the Gold Rush look tame. Harvesting took place from islands off Namibia, Oman, Patagonia and California – mirroring the industry that blossomed off Peru soon after von Liebig's treatise. In 1856, the US passed the Guano Islands Act which gave US citizens who discovered guano on unclaimed islands the exclusive rights to deposits (interestingly, nine of these islands are still officially US territories). China also got in on the act of guano-chasing. What started in 1849 as 80 or so Chinese guano-harvesters turned into an industry involving 100,000 of their countrymen two decades later. Predictably, wars were fought over who had control over the most important islands, namely the Chincha Islands War (1864–1866), where Spanish soldiers fought a Peruvian–Chilean alliance, and then the subsequent War of the Pacific (1879–1883), in which Chile seized many of Peru's guano riches. Apparently, so rich were sales of the faecal treasure that taxes coming from these newly acquired lands saw Chile's national treasury increase by 900 per cent by 1902.

Other countries wanted in, not least South Africa, which began to independently harvest its own guano-filled islands, to the detriment of the creature with which this chapter began: the African penguin. African penguins make their nests from guano. They make burrows in the stuff into which their eggs are laid. With less guano going spare because of human extraction, penguins began making nests in the open, exposed to the elements. It surprised no one when their numbers plummeted. In stepped the conservationists and a captive-breeding programme began, sending penguins to far-flung places like Torquay to breed in the hope that one day they might be reintroduced back into an ocean environment now overfished to within an inch of its life and missing all of its guano to boot. So, in a nutshell, this is why there was an African penguin in that river on the edge of Azerbaijan. Our lust for guano started it all.

I had debated whether or not this book needed an exploration of faeces. I'm still unsure, to be honest, but I think it is justified because faeces is, essentially, death. It is a product of life. It is where dead things go through living things; it is arguably the single biggest currency of death that we see in day-to-day life. And, like dead bodies, it is life-giving (which is why the global trade in the stuff emerged in the first place). So allow me a few more moments to talk about guano, if I may. And then we'll be done with it.

The best and most nutrient-rich guano comes from islands that don't receive much rain, where the sparse rainwater leaches the nitrogen-containing ammonia out of the guano but fails to wash it away. The truly best guano often comes from islands next to oceanic upwellings where nutrients surge up from the depths and are feasted upon by fish, which are then feasted upon by seabirds, who then squirt the nutrients from the upwelling onto the rocks in shiny wet new faecal form where they remain until harvested. Many landscapes and seascapes are revitalised by this material that we might be tempted to call 'waste' – in reality it is anything but. It really is life-giving.

Perhaps the best-known example of how valuable to life faeces can be is that of the 'whale pump' – an idea put forward by the marine biologists Joe Roman and James McCarthy in 2010. The whale pump is a system of nutrient transfer from the ocean depths (where whales may feed) into the surface waters where they surface to breathe (and where whales defecate). You or I might picture something heavy, big and brown coming from a whale that sinks quickly into the depths, but that's not the case with whales. Their faeces are known as 'flocculent faecal plumes' – they squirt forth from the whale's anus and then dissipate into a cloudy layer that sits in the surface waters. This layer becomes like newly fertilised soil to the planktonic organisms that dwell there. Life moves in and flourishes using the nitrogen that exists there in huge concentrations.

'Whales and seals may be responsible for replenishing 2.3×10^4 metric tons of N [nitrogen] per year in the Gulf of Maine's euphotic zone, more than the input of all rivers combined,' write the authors in their 2010 paper in *PLOS ONE*. And that's after centuries of brutal treatment at the hands of humanity. It's giddying to imagine what an ancestral ocean of whales and seals could have achieved in terms of fertilising the ocean surface. More nutrient movement than the guano industry in its prime, certainly. Whaling nations sometimes argue that whales compete indirectly for fish. Well, if Roman and McCarthy are right, it's probably the opposite way round. Whales seed the surface waters with nitrogen. They bring up nitrogen from the depths and create a habitat in which young fish flourish and adult fish feed. And we have the gall to blame them for our reduced fish stocks? No wonder sperm whales have been known to fire massive shit-clouds at passing divers.

Bats are another famous producer of highly sought-after guano. In many caves there are great mounds of the stuff, accumulated over thousands of years. In these unusual habitats, an ecosystem has blossomed that is home to hundreds of guano-adapted invertebrates that include nematode worms, millipedes, mites, springtails, micro-moths and flies of numerous species (there is even a guano-eating amphibian, the aptly named grotto salamander). Mites are especially common. In one New South Wales cave there were found to be something like 12.6 million mites per square metre of guano. As I had learned through the pig corpse, great swathes of predators join the guano-induced party. Rove beetles, ground beetles, pseudoscorpions, centipedes, harvestmen, spiders, moths and parasitic wasps are only some – they feed on the mites, the springtails and the nematodes, particularly. What's perhaps most impressive is how the sex lives of all of these creatures are largely governed by the abundance of bats and droppings in any given month. Even in a cave, without light, there are seasons brought to them by the frequency with which the bats above them produce their faeces (frequent droppings = summer; more infrequent droppings = winter). Life is made of wonderful phenomena like this.

Sadly, as with seabird islands, bat guano is often too good for humans to let sit. Far better, says humankind, to mine it. To sell it. To ship it. And so they have done just this. Bats don't particularly like it when people with machinery enter their caves. They do that thing that bats do in movies where it looks like they're attacking *en masse* but the reality is that they're all panicking wildly and flailing around not really knowing what to do. What you don't see in films is that some of these bats simply expire and fall to the floor dead during these human intrusions. Many young pups will also die, accidentally dropped by their mothers into the great swarming mass of life waiting beneath. Its likely that many bat populations have been damaged by guano-mining. And with the bats goes the guano. And with the guano go 12.6 million mites per square metre. What a sad loss of life. What a sad loss of lives …

Thankfully, though (depending on how much of a fan of mites you are), guano doesn't have quite the market price it once had and so the industry of guano collection isn't quite as relentlessly barbaric as it once was. In 1909 Fritz Haber came up with the Haber–Bosch process of industrial nitrogen fixation, which now generates enough ammonia-based fertiliser to sustain an estimated one-third of all humans on Earth. Trillions of faecal mites owe their continued existence to this man. Millions of centipedes, too. Still, it hasn't been a happy ending for all guano-loving creatures. Many African penguins remain holed up in captivity; many African penguins aren't where they should be. Yet. Hopefully one day they can return to their island havens, stinking and covered, as they should be, in a thick layer of faeces provided like mother's milk by previous generations.

This has only been a short chapter. I wondered whether to make it longer. I thought about focusing more heavily on faeces, but others have gone there and done a much better … job.* At one point I even considered visiting a sewage treatment works – I made contact with my local sewage treatment centre to find out more about what, exactly, happens to my faeces and the faeces of my toilet-trained children when we flushed the chain. I was surprised to discover that, actually, they weren't enormously interested in me 'popping in', which was a shame because I pay them something like £500 a year for this service. You'd think they'd be pulling people off the street to see what they do with all that money. But no. They weren't really kitted out for visitors, I was told.

And so thankfully (for you) I chose not to chase up this particular thread. I still think this is a bit of a shame. I think we might have a better appreciation of what life is all about if we could see what happens to our faeces. It is the one thing that bonds us, after all: tube-shaped animals with a mouth at one end and an anus at the other. A mouth into which death passes; an anus from which death spews and onto which life returns, if and when we allow it; if we cherish it enough and leave it where it lies as best possible. For it is from shit that some good things come. It is from its faecal plumes that some life blossoms, even if it's just a collection of centipedes and mites. It's still life. And life is, on the whole, quite wonderful.

*I recommend *The Origin of Feces* by David Waltner-Toews.

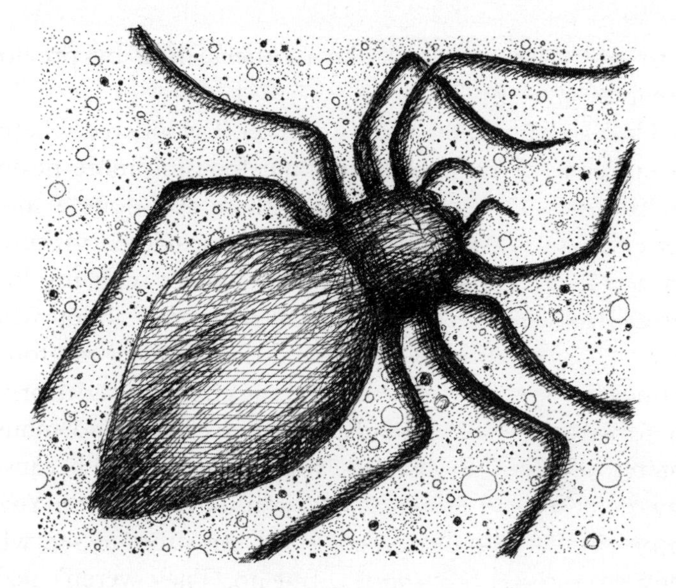

CHAPTER TWELVE

The Horrid Ground-weaver

A security guard strides over to us from across the car park looking notably disgruntled, as if his time is already being wasted just by approaching us. I can understand why he's come over, mind you. He is wondering why three people are standing in a car park outside his hardware store looking up and staring at the cliff-face that overhangs it. As he approaches he possibly wonders why one of us is wearing a lime-green t-shirt with a big picture of a house fly on the front. He's thinking: 'Who stands around in a car park and wears a t-shirt with a big picture of a house fly on the front?' He is no doubt wondering also why one of us is playing with an incredibly complicated handheld GPS device, and what this could mean. He may also be confused about my role – a man nodding his head too much, waving his arms a lot and recording the whole thing with a dictaphone.

'Can I help you?' he asks when he reaches us. Andrew Whitehouse, in the house-fly t-shirt, spins around from the cliff-face and faces the security guard confidently. He speaks as if it's the most natural thing in the world to stare excitedly at a cliff, like this is all totally normal, but I think that there's a hint of self-doubt in there too. There's no getting around this. Andrew is, after all, about to ask permission from this man to look for a tiny money spider that may or may not live on his business's property.

'Do you know who manages and looks after this site?' Andrew asks the security guard. 'Erm,' says the guard, suddenly caught off-guard. 'I haven't got the details to hand, but …' The security guard regains his composure. He almost shakes himself back into reality. 'Wait,' he says. 'Who are you? What are you doing here?' Ah. Ah, right. We all look slightly embarrassed at the security guard's sudden grasping of authority. I shuffle around slightly. 'Are you birdwatchers?' the man says, suddenly quite serious. 'Well,' begins Andrew. '… no, we're not birdwatchers … It's worse even than that.' He takes a deep breath, gathering his thoughts before continuing. The security guard crosses his arms. 'Ok,' Andrew says. 'So, here's what's going on: a really rare spider was discovered in this quarry before this retail park was built and before your business moved in here. It's called the horrid ground-weaver.' Andrew pauses. 'Right,' says the security official, deadpan. He nods his head imperceptibly slowly. 'Go on …' Andrew takes his cue. 'It's called the horrid ground-weaver,' he says again. 'It's only known from three sites in Plymouth. Plymouth is the only place on the planet that the horrid ground-weaver exists. And it used to live here, on these cliffs, when this place was a quarry.' Andrew looks up at the limestone cliffs and the assembled boulders that jut out from the sandy soil beneath. The security guard's eyes remain firmly on

Andrew. He looks really quite serious. Andrew catches his look and stands firm, determined not to lose his composure. 'We need to know if it's still here,' Andrew says confidently to the guard. There is a pause. 'Ok …' says the security guard, slightly more suspiciously than before. 'Ok …' he says again. Being honest, this isn't going quite as well as we'd hoped.

There are a few moments of silence again. Andrew tries a new approach: a child-like grin. 'This sounds crazy. I know it sounds crazy, but …' He draws breath. 'What we'd actually quite like to do is get up on that cliff, with all the right safety gear, and look for the spider … actually get up there and look, you know?' he says, before adding hurriedly, 'But first, of course, we need to talk to you.' He says this last bit with a big open smile again. The security guard looks at us all. A pause. 'Where are you from again?' he asks. 'We're from Buglife,' says Andrew, as if Buglife is some sort of well-known law-enforcement agency (I half expect him to pull out a badge). There is one final silence. 'Ok,' says the security guard with a slight smile. 'You'd better come with me to speak to the boss.' Andrew walks off with the security guard to the main building, leaving his colleague Jo Gilvear and I standing alone in the car park. Before he goes in Andrew looks back at us, offering us a little thumbs up. He enters the building.

The horrid ground-weaver: a money spider so nondescript they had to give it a funny name so everyone would remember it. So nondescript that, to correctly identify it, we'd have to kill it and examine its genitals under a microscope. So nondescript it eluded invertebrate experts for centuries, until finally being identified as its own species (within its own genus, no less) in 1995. I was in Plymouth hoping to come face-to-face with perhaps one of the rarest animals on the planet.

I got here through John, the journalist responsible for the false widow hysteria. Hearing his voice in that coffee shop, knowing the hatred that he had unthinkingly stirred up against the nation's spiders, I had suddenly felt it my duty to stand up for spiders even more in recent months, being that they are largely quite lovely, at least on the whole. Even though I'm a tiny bit phobic, spiders really are amazing. I can see that quite clearly now. Spiders are an invertebrate class that hit upon a design so perfect for predation that they have cemented themselves into food webs the world over for hundreds of millions of years.

The plight of the threatened horrid ground-weaver in recent months has had me thoroughly ensnared, being that it has largely played out in the public arena, something that doesn't often happen with inconsequentially small and quite boring money spiders. The reason for this public media attention is that the horrid ground-weaver is a spider species that humans are debating killing totally, knowingly forcing it to extinction, in the name of 50 new houses. Sure, there's a small chance it might still exist on the cliff next to the car park where we had just gathered, but it probably doesn't. Its stronghold is thought to be two miles away at a place called Radford Quarry, a former industrial site that has since become a haven for wildlife, and which is now threatened by becoming a housing estate. I use the term 'stronghold' loosely. Really I mean: 'place where it has been seen five times'. Either way, that place, the former quarry, is now threatened and Buglife are calling for it to be saved. A spider, the only one of its genus in the world, which has only been spotted a handful of times ever, is about to have its fate decided by us. By humans. What can I say. The story really appealed to me. It had drawn me in.

Whilst writing this book, I have tried to define life and death, and looked at the death niche and the invertebrates and vertebrates associated with it. But I'd also seen something I hadn't expected: humans killing trees in the mistaken view that caterpillars were monstrous, humans generating advertising revenue from spider scare stories, humans needlessly killing and demonising scavengers. And now, here I was in the middle of another strange human behaviour surrounding death. Here I was in the middle of a pack of humans debating and warring over whether, and to what degree, the total extinction of a money spider really matters to anyone. And the debate was actually becoming incredibly interesting. I mean, consider it for a moment: the history of Earth is littered with extinctions, so why the bloody hell should it matter if a tiny spider lives or dies? But then there was the opposition, the conservationists questioning how we might live with ourselves knowing that we had chosen to cause a total extinction of a species. What's going on with them? How did they get so moral about it? It has been entrancing to watch the story of the horrid ground-weaver play out through the press. Gripping. Really gripping.

So should it matter if we let species go extinct? Contrary to what you might think, I haven't really cemented my thoughts on the issue. I'm willing to have my opinions challenged, I guess. The potential extinction of a tiny spider with slightly hairy legs that lives underneath limestone boulders and that only a handful of arachnologists has ever seen seems a good testing ground for the moral argument for or against extinction, I suppose. Like I said, I was drawn to it. In many ways, our reaction was confusing, partly because extinction – the total death of a species – is so very common in Earth's history. Who cares if we cause it?

Famously, it's said that 99 per cent of all species that have ever lived are now extinct. The death of species really is, like the blossoming of new species, *de rigueur* in the history of life on Earth. Most notably these die-offs occur during times of mass extinctions, such as that which occurred 252 million years ago (when 95 per cent of marine invertebrates were made extinct) and 65 million years ago when most dinosaurs died out, paving the way for the slow advance of aye-ayes and aardvarks and other nippled creatures. It's becoming increasingly clear that we live in a time of similar such extinction now, possibly at the dawn of another mass extinction – a mass extinction potentially like no other given that, this time, a single creature is the cause. Us. Of the 44,838 species categorised by the IUCN under their Red List criteria, 16,928 (38 per cent) are currently threatened with extinction. Our oceans are dying, we're told. Our rainforests are pillaged. Our grasslands are becoming deserts. According to WWF, the Earth lost half of its wild animal populations in the last 40 years. Half. In the UK, 60 per cent of native species are declining and recent research says that 1 in 10 are on their way to national extinction. These are terrible times to be anything other than a human, a chicken, a cow, a sheep or a crop monoculture.

Yes, extinction really is commonplace. The tree of life on Earth is littered with branches and twigs upon which leaves no longer grow. Though the great boughs stand strong, whole branches have withered and died, or left twigs upon which only a few leaves grow today. In our own primate lineage, fossils show us that our ancestors were once part of a well-twigged and well-leafed branch that has withered greatly. Apes particularly were at one time quite a diverse and funky lot. Now their numbers are sliding precariously downwards; in many ways those that remain on Earth are a shadow of their former diversity. It's a similar story with the horrid ground-weaver – it is a spider that sits within a genus that was probably once thriving and full of life, but that now numbers only a single species. A single leaf on a long-forgotten twig in the tree of life. Possibly, or quite likely, on its way out anyway. Extinct. So why does it really matter?

Jo and I kick our heels in the car park while we wait for Andrew inside, speaking to the manager about whether he's happy with a team of strangers in weird house-fly t-shirts ascending a cliff on his property, potentially risking life and death for a spider many of us wouldn't think twice about squashing with our thumb. Neither I nor Jo are quite sure whether Andrew will come out of the building victorious or with his head down and his shoulders slumped. We make small talk. Jo is a freshwater biologist who is starting a new(ish) life as an arachnologist. It will be her job to manage Buglife's (largely crowd-funded) Horrid Ground-Weaver Project, which it's hoped will find the spider still alive on the nearby disused quarry or here, where it was once spotted two decades before on the cliffs. Should that happen the next step ideally would be to get the spider protected by law and, even more ideally, protected from encroaching housing developments, like those that currently loom.

Jo is excited but slightly daunted at the prospect of raising awareness about what is genuinely a fairly nondescript money spider. 'So tell me,' I say. 'Why is it worth saving?' I deliver the question quite nonchalantly, pretending it's not the only reason I'm there. 'Why does it matter, exactly?' I'm not very good at this style of direct questioning but it seems like an important place to start; a way to figure out why the spider has captivated local people so much. Jo is like lightning with her response. She says it kindly and with feeling and genuine heart, but she does seem to bristle slightly. 'Why would anything going extinct *matter*?' she says quite abruptly. She gathers herself. 'For me … it's a moral issue. I mean, we've found it. We've found this spider. We know it's almost gone. As soon as we realised that it existed it's become our responsibility. We had to act. Someone has to act. That's why we've got to save it. That's why we're here.' Her voice trails off a little, almost like she's embarrassed that she allowed herself to become so animated. 'I guess we don't know much about this spider,' she continues, more slowly this time. We stand for a few moments in total silence. She looks back to see if Andrew has popped out of the door yet. She looks back at me and then at the floor. 'How sad would it be if this thing could come and go without us ever knowing much about it, or adequately documenting it? How sad would it be for this spider to be this … this … this fleeting thing?' she says.

A fleeting thing, like all things. But a fleeting thing that is facing extinction on our watch. Because of us. I agree with Jo that this makes it different, somehow. I stand there thinking it over. If the arachnologist who first stumbled upon this creature had never thought to identify the tiny spider he'd found, we would never have known it existed. We would never have been forced to act. And it is probably one of only a handful of spiders to ever have received conservation funds to save it. It is certainly the first spider ever that experts are attempting to save through crowd-funding, we can be sure about that. But the question remains: why save it? Why does it matter?

Much of my experience with conservation, as I mentioned earlier in this book, is with amphibians. Amphibians are animals that will always be close to my heart because they are the taxonomic class with which I began my career. And this is a vertebrate class that really is in trouble. In 2008, one study calculated that the current rate of amphibian extinction could be more than 200 times greater than the background extinction rate. This is not good. A key part of my role in that first job (part of which involved manning the infamous frog helpline) was to let people know why amphibians are worth saving. I had to convince people (mainly potential funders and donors) that we shouldn't just let them die, and that we'd need their cash to help save them. But how do you convince people to want to save frogs? Simple, I was told. I'd had the answers to this question drilled into me when I first started the job. The reasons amphibians were worth saving, I was told, were threefold:

1. They eat pests.
2. The freshwater habitats in which they breed are crucial for human life and livelihoods.
3. They might be a source of new medicines and painkillers.

And that was it. Three reasons. That was all it would take, my bosses hoped, for me to convince the public that frogs were worth saving. Three reasons would be all it would take for punters to put their hands into their pockets and pull out wads of cash. I did as they said. I brought up these three reasons why amphibians were worth saving, just as instructed. But the more I spouted out these three reasons for why frogs mattered, the more I hated how they sounded. For a start, I couldn't help this feeling that they were such awful and tawdry reasons. So arrogant and human-centric, somehow. I wished I was confident enough to have screamed something else. Something like: 'WHY SAVE AMPHIBIANS? REALLY? WHY??? WELL THEY'RE SO WEIRD FOR A BLOODY START. THEY ARE VESTIGES OF FISH-LIKE CREATURES THAT NEVER NEEDED TO EVOLVE SHELLED EGGS! LONG-LEGGED! LONG-LIVED! AN INSPIRATION TO JIM HENSON! AN INSPIRATION TO NATIONS! NIGHT CALLERS! NIGHT STALKERS! MASTER NAVIGATORS! EXPLOSIVE SEXUAL DYNAMOS! BODY HUGGERS! MAGGOT CHUGGERS! FEEL-GOOD DAY-GLO PO-FACED PRIMITIVES THAT DESERVE NO SUCH TITLE BECAUSE THEY'RE STILL HERE AND SO ARE WE SO, YES, WE SHOULD BLOODY SAVE THEM BECAUSE IF WE DON'T WE ARE A DISGRACE AND HOW WILL WE LOOK INTO THE EYES OF OUR CHILDREN?' *etc. etc.*

I really wish I could have said that. Frogs and toads and newts are fantastic in all sort of ways. They are clearly weird. They are clearly quirky. And they're about much more than what we can take from them. But, that's frogs for you … so what about spiders? Why would anyone want to save them? A tiny nothing-much of an invertebrate; who'd want to save that? They're not particularly quirky or enigmatic or charismatic. They're just … tiny spiders. But people really did seem motivated to act; people really were motivated to try to save this tiny spider from total destruction.

When news of the horrid ground-weaver's fate had spread, Buglife launched a petition to influence the planning inspector responsible for considering whether or not the housing estate proposed for Radford Quarry could or could not be permitted. It encouraged the planning inspector to halt the proposal in its tracks; to ditch the housing plans for the sake of the spiders. Of the 9,732 signatures the petition received within a week, many signatories took a few moments to add their own reasons why they felt this little spider was worth saving. There was a little comment box on the petition web page especially for these people. I read through these comments with deep interest. Could they shed light on why people thought that the tiny spiders were worth saving? Maybe. The first read: 'Because it shouldn't matter if it's a panda, a clouded leopard or a tiny obscure spider, vulnerable endemic species are all equally important to the world.' I thought about this. Maybe that's true? I wondered. 'I grew up round here,' says the next comment. 'The spiders are my friends.' 'Can't believe this is in my home town!' says another. 'No doubt these property developers are going to appeal again and again. Let's hope they don't get their way.'

These were interesting reasons to save the horrid ground-weaver and not necessarily what I had expected to read. Not one of the hundreds of comments talked about what we humans might gain from their survival. Not one of the comments talked about cures for insomnia or erectile dysfunction (which is genuinely something some spider venom might cure, for those interested). The arguments that they gave were all moral ones. Almost every one was about love and feelings and emotions. The spiders needed saving 'because every species deserves protection not just the cute and large mammals'. They needed saving because 'we have few endemic species in the UK we should all be fighting to save it not standing by and letting another extinction go unnoticed'. How strange then that as a young conservationist in my first job on that frog helpline I overlooked this moral argument so readily. I never thought that people could be asked to do something or donate to something to be moral. To be better people. How had I so easily missed this?

Andrew comes out of the trade shop. He is smiling. He almost, but not quite, struts toward us. 'Sorry about that guys,' he says smoothly as he gets closer to us. 'I've been trying to crack speaking to them for quite a while. That was great. All done now. Got the contact details of the landlord. Hopefully we'll be good to go.' Jo and I make approving congratulatory noises. Soon Andrew and Jo will return with hard hats and climbing gear. In time they will hopefully be able to better gauge whether or not the horrid ground-weaver still remains here at this site.

We walk back to my tiny red car, scouring the cliff edge with our eyes as we walk, wondering how many horrid ground-weavers there might be scuttling among the rocks. I unlock the car doors and we get in. The plan is to head over to Radford Quarry, which is what most consider to probably be the spider's major stronghold. Andrew directs us out of the retail estate, then up a hill and into a dense new-build maze of streets and houses with new cars parked in front of the drives and on the pavements and on the pavements over the road. (Many new housing estates in Britain seem to have this problem – they seem to dramatically misjudge how many cars most families have. It's two. Two.) Going left and right, packed into the little car like sardines, we bump and bustle into each other as we weave down more busy streets and over what feels like hundreds of pedestrian crossings. Cars squeeze past us. Vans. Lorries. We climb higher across Plymouth, until we're almost looking out over the bay. More streets. More winding. And then we're there. We park up next to a small gravel trackway at the edge of a different new housing estate, and I open the doors and breathe in the industrial fug coming in off the docklands.

Andrew leads the way to Radford Quarry. We start walking down the thin gravel track. We are not the only ones to have trodden this path – there are bike tracks. The path is well trodden. For reasons I have often struggled to fathom, someone has left a dog poo in a blue plastic bag hanging off an overhanging branch. Andrew and Jo don't even seem to notice. Andrew tells me that what this place represents is a remnant kind of limestone and grassland habitat that would have once extended across this whole area and all the way across Plymouth's industrial centres. Radford Quarry is one of the largest patches of exposed limestone that remains locally from this former age. He explains how the Planning Inspectorate are still deciding, and have been for months, whether to permit 50 houses to be built on the site. He talks about how upset the local people have got; how interest has grown in the little spider. If the houses get built, the spiders here will surely die, he explains bleakly.

We walk down a concrete path through thick vegetation and come to a large muddy path, the entrance into the disused quarry. Again, the pathway looks well trodden by pushbikes, motorbikes and dog-walkers. We stop at the entrance. I wonder why we are stopping. Why aren't we going in? At this point Andrew politely points out that there is no way we can go in there (even though lots of people clearly have done just this) because it's not legally public access. Oh, I think. He and his colleagues aren't really allowed on the site and nor am I. No one is, legally. We'd be trespassing, he says. Ah, I think. A small part of me wonders whether if Buglife was allowed regular access to the site they would find all manner of other unusual invertebrates, which would probably make future proposals for new housing estates here that bit more difficult. I choose not to raise this with Andrew and Jo. I am a tiny bit downcast at not being able to enter, though.

'It's private property, so that's that,' Andrew says, resigned. 'But we are hoping we will be allowed further access to the site to survey at some point.' I look longingly up the path and into Radford Quarry, imagining all the spiders that only a few people in history have ever seen alive.

Radford Quarry (if only by looking at the entrance) didn't quite look like the natural place that I had imagined. From where we stood there were about 2,000 houses just out of view on the hillside. Busy streets. Delivery drivers rushing here and there. Traffic lights. Road markings like graffiti. School runs. The former quarry looked to be a pocket of life within all of this activity. There was so much green in there, plenty of saplings and grass. It seemed to be the only place around here that had gone from an industrial slate-grey to a warm green, rather than the other way around. In Britain we call former industrial places like this quarry 'brownfields' as if that's all they are: brown fields. I hate this label. I deeply hate it. It's so shockingly small-minded to describe them so blandly, like hearing someone describing coral as 'pointy sea-rocks' or plants as 'those green things'. Ill-judged. Patronising. Yes, small-minded. Brown? I sigh.

I peel my eyes off the entrance to Radford Quarry and look down at the stones and the green grassy verges to the path we had just taken from the parked car. 'Do you think we might find any spiders outside the entrance, around here?' I ask Andrew and Jo. 'Almost definitely not, but we could have a look for some other invertebrates ...' says Andrew, smiling.

In modern conservation, journalists covering stories like the horrid ground-weaver might go for a simple narrative, classically encapsulated as extinction being a tale of 'us versus them'. Of David and Goliath. I guess it's the most grabby way of telling the story, but in many ways this is too simple a view. Extinction is very rarely a two-dimensional war between individual species. It's an oversimplification to talk of it as a battle of occupancy between one species (normally us) and another species (like horrid ground-weavers); it's much more complex than that. Breathtakingly so, in fact. In reality, extinction is complex and multifarious, and it demands of us a different mental picture to grasp it properly, one originally constructed by, of all people, Charles Darwin.

Long before publishing *On the Origin of Species* Darwin wrestled with his (then vague) notion of natural selection, and how it operated in nature. His idea was simple, sure – a series of successful copying mistakes is not hard to apply in one's mind – but Darwin knew that the concept, in this form, was too linear. It wouldn't work for Darwin. He needed more. Darwin knew that the world is anything but linear. It's complicated. It's messy. Darwin needed a metaphor to graphically communicate his big idea to his audiences, but more importantly he needed a metaphor to toy with, so he could pull apart what his idea looked like, exactly, when applied to the natural world. On 28th September 1838 he realised it. After reading Thomas Malthus's *Essay on the Principles of Population* he scribbled the following into his notepad: 'One may say there is a force like a hundred thousand wedges trying [to] force every kind of adapted structure into the gaps in the economy of nature, or rather forming gaps by thrusting out weaker ones.'

It sounds simple, but this was a rather enormous insight to have. Indeed, many Darwin scholars (including Stephen Jay Gould in his wonderful essay *The Wheel of Fortune and the Wedge of Progress*) think it was his biggest. Gould, in particular, argued that it was this metaphor that drove his theory of evolution by natural selection forward to publication 20 years later. For in that metaphor is everything one needs to know about the death of life's lineages and what extinction actually looks and feels like. Darwin later honed the metaphor further. In his manuscript for the longer unpublished version of *On the Origin of Species* his metaphor became more elaborate and even more delicious. He wrote:

> *Nature may be compared to a surface covered with ten thousand sharp wedges, many of the same shape and many of different shapes representing different species, all packed closely together and all driven in by incessant blows: the blows being far severer at one time than at another; sometimes a wedge of one form and sometimes another being struck; the one driven deeply in forcing out others; with the jar and shock often transmitted very far to other wedges in many lines of direction.*

And that was what I thought of when I was on my hands and knees looking for that bloody spider, aware of all the houses and the hustle and bustle of city life that surrounded us. The horrid ground-weaver is the tiniest of wedges in a writhing nebulous wobbly mass of a million wedges. It is one of the tiniest of nearly all wedges in nature; occupying a simple, simple, *simple* niche: that of cracks between limestone rocks in a limestone quarry in Plymouth. The human wedge (which is being struck particularly hard around here) is upsetting this tiny wedge, threatening to squirt it out of the pack and into oblivion. But in an ecological sense, the spider probably doesn't really matter much at all. Sure, there might be a period of rejigging, but the wedges of other species won't collapse upon one another. The other wedges will probably not even shift should the spider indeed slip toward extinction. And so, to nature, the horrid ground-weaver is instantly forgettable. Darwin understood that extinction, as well as being a very real part of the fate of many creatures, is also a measure of the squeeze in nature. A squeeze ultimately limited by how much there is to go around. And this is probably why, in some perverse way, I'm quite sad at the thought of not getting to see this spider. I had wanted to see it very much.

I had wanted to hold one. To give it new value – a human value – by holding it in my hands and by looking at it face-to-face: wedge-to-wedge.

Sadly, though, we fail to find a horrid ground-weaver. We see plenty of pillbugs (a kind of woodlouse that can roll into a ball). Thousands upon thousands of them. They are the shiniest and most pristine pillbugs I have ever seen in my life; it looks like someone has been here just before us to polish them. Each looks resplendent. There are bristletails, too, prehistoric-looking creatures that look a little like miniature horseshoe crabs with long tail feathers. Then there are the jumping spiders, which hop and scurry out of view with every step we take, and the velvet mites, the wolf spiders, the pseudoscorpions. Occasional peacock butterflies, taking a few moments to sun themselves upon the grassy banks of the path. There are also millipedes, ground beetles, flower beetles, rove beetles and midges; rich ecosystems written in fonts almost too small to read. But, alas, no horrid ground-weaver.

As we walk up and down the path outside of the quarry, Jo informs me that her plan over the next few weeks, if all goes well, is to build bridges with the landowners to gain access to Radford Quarry to survey for the horrid ground-weaver. Her aim is to get out here with bug-vacs (read: hoovers) and to set up pitfall traps to look for the spider. There's even a possibility, she tells me, that later in the year local students will be encouraged to gather and do a full hand-sweep of the area like police officers scouring a crime scene. 'We'll find it,' says Andrew confidently from up the path, hearing Jo speak. I have visions of him as an old man, still traipsing around this place with his bug-vac saying rosily 'we'll find it again …' after 50 years of searching.

And so I have to accept that the horrid ground-weaver has eluded us. I won't get my moment, I realise. As we stroll back to the car I find myself confessing to Andrew and Jo about how troubled I have become with the idea of why, in a world of death and extinction, we should become so bothered about conserving all species, including tiny spiders, into perpetuity. Andrew is pleased that I brought it up, I think. 'Yes, the horrid ground-weaver isn't going to blow people away like seeing a blue whale from a boat,' he said, half smiling. 'But every species has an equal right to live on this planet, no matter which way you look at it.' Jo reiterated what she had said earlier: 'We can't knowingly let it slip away now that we know it was here. To do so would be wrong, somehow …' I realised that Andrew and Jo were actually very moral people, but not at all in a preachy way.

Andrew got a little quiet from this point onwards. 'Extinction is wrong,' he said as we neared the car. He was suddenly very serious. I asked again why this little spider mattered so much to him. 'It's not going to change anyone's life if this thing goes extinct, I agree with that statement,' he said, thinking it over. And then he stopped and turned to us both. 'Except for mine,' he said sadly. 'My life would change if we let it become extinct, because I'll feel sad that we let it go. That I'd been part of its extinction. I'd be gutted. I'd be gutted and frustrated. I would just be so sad.' We headed home.

If conservation is about the moral imperative, surely Andrew and Jo are taking part in one of the most moral acts ever witnessed by humankind. They are saving something not for our children to see; saving something not for medicinal gain; saving something not for the value it has as a controller of pests or pestilence; saving something not for its quirkiness or the special skills it possesses. Nope, none of these things. They are saving an animal which has no cognition or knowledge of its own existence in any way. It literally has no purpose other to make more horrid ground-weavers, and even its ability to do that is questionable. Yet still they want to save it. And incredibly, Andrew and Jo have never even seen a live one. They've never seen or held or watched up close a living horrid ground-weaver in the flesh. Yet they care.

Somehow, against all odds, a small (and growing) group of entomologists and amateur enthusiasts really do care about it.

There is no meaning in nature. Unless you give it meaning, that is. And extinction doesn't always matter, unless you think it matters. Unless you give it meaning, I mean. We are the first animals on Earth to have done this. To have given extinction meaning. We can't be all that bad, then … can we?

CHAPTER THIRTEEN

Dark Matters

Was I getting a bit weird about death? Was all this talk of maggots and disgust and extinction and money spiders warping my mind? I was starting to worry. I was well over a year into my journey. I decided I needed to seek professional help. Help from professional zoologists. I wanted to know how they dealt with the relentlessness of it all, the endless thinking about death. Of the handful of zoologists I approached about this, one replied almost immediately. She was Anne Hilborn, a cheetah ecologist based at Virginia Tech.

I had followed Anne for many months on Twitter, partly because the habitat in which she bases her studies seems so tough and so brutally *different* to the professional world which I inhabit. She spends much of her time in the Serengeti, and one of her study animals is the cheetah. The photos that she posts on Twitter are particularly and fantastically real: starving cheetah adults, lions with ticks under their eyes, hyenas covered top-to-tail in swarms of biting flies. That sort of thing. The world in which she bases her studies seems so stark that, in some ways, it's a breath of fresh air to me. I sensed Anne would be worth approaching. I sensed that, given her line of work, Anne thinks a great deal about death given that, in cheetahs, that's pretty much most of what she seems to see. Dead baby cheetahs. And there's a reason for this. In cheetahs, juvenile mortality rates are staggering. They appear to die far more than anyone had ever predicted before they thought to look. From a three-year cheetah tracking study in the nineties (undertaken and led by the ecologist Karen Laurenson) it was found that for every cheetah born, only 5 per cent – only 1 out of 20 young cheetahs – survived to adulthood. This had been a real surprise. The other 95 per cent of young cheetahs were found to die in lots of ways. Many were victims of starvation or abandoned by their mothers during forest fires or incremental weather events. Most were killed by lions or spotted hyenas (though strangely, they were often left uneaten).

The loss amazed me when I first heard about it. I'm used to such statistics in frogs (where each blob of frogspawn might have 1,000 or more eggs with only three or four hatchlings making it to adulthood). But these were cheetahs. Big bloody cats. What the hell was nature up to? If animals mourn, then cheetah mothers must spend their lives in constant distress and turmoil. How do the female cheetahs deal with all the loss? This was the first question I put to Anne and her response was suitably rigorous. 'Very few of our cheetahs are seen regularly so we almost never see cub death or its immediate aftermath,' she said. 'I think I have seen pictures of a mother cheetah carrying her small cubs that had been killed by lions, but I can't remember the details … I assume it was for only a very short amount of time. I am not sure if that qualifies as mourning or not. It's interesting to think about how or when cheetahs might accept or recognise that a member of their social group who has been missing is actually dead. I don't have any definite observation or data on it.' Cheetahs appear so often on TV documentaries, I had kind of assumed we knew everything about them. But, according to Anne, many parts of their lives still seem very secretive. They are hard animals to get to know, it seemed.

What I liked about Anne was that she felt free and easy talking about death on the savanna; it had almost become second nature to her. And she talked personally about how it had affected her own attitude to death as a result. 'As a child I loved animals in a very sentimental way,' she wrote to me. 'Over years of fieldwork on various animals my feelings have changed. Doing fieldwork in Alaska on sockeye salmon you see a lot of carnage. We work on really small streams that for a week or so are crammed with salmon trying to spawn. These streams are so small, sometimes the water does not cover the backs of the fish and they are ridiculously easy prey, not only for bears but also for gulls,' she wrote. 'Gulls go for the eyes, and will pick them out of live salmon. And salmon can survive for quite a while without eyes. I find something particularly horrifying about having your eyes pecked out, so I always tried to kill live fish who'd lost their eyes (with a swift knife slice through the brain). We'd also see live salmon that had been mauled by bears still swimming around with bites out of their backs, or slash marks on their sides. Seeing this sort of thing daily does "toughen" you up, and I've gotten pretty good at not feeling much emotion at seeing things die.'

What Anne said next really interested me, because it is almost exactly how I have come to feel about death whilst researching this book. 'Sometimes I wonder if being toughened to death is a good thing. I think it makes me a better biologist and it allows me to do the fieldwork and research I love. But occasionally I think that maybe I've gone too far, and my lack of feeling at the sight of death means I have become callous, and that people who are distressed at the sight of gruesome pictures might be right.' This was me in a nutshell. Recently I seemed to be spending a great deal of my time worrying about whether I'm starting to come across as an Ultra Naturalist; a life- and death-obsessed person with whom normal people can no longer connect. Death is so inherent to life, it's becoming quite natural to me to introduce it into everyday conversations. I try to hold it back, but sometimes it slips out. Over coffee with a friend, I might mention that I recently stood in a field of dead pigs or held a 500-year-old shellfish or something like that. It's almost out of my control now. I worry that I'm being judged for it.

'Not all biologists are the same,' Anne wrote. 'I've worked with people who laughed and joked about killing salmon to take blood samples, and with people who thought my fascination with bones and dead things was morbid and distasteful. But in general I would say that watching a lot of death in the natural world does tend to harden people and strips away sentimentality about death and any 'Disneyfied' ideas about animals living in 'harmony' or 'peace'. As far as I can tell most animals die in a manner humans consider gruesome. Either they are eaten by predators (often alive), killed by conspecifics, starve to death, or die of some nasty disease. I know of very few 'easy' or 'happy' deaths in the natural world. A quick death by a predator is probably the death that would be considered 'best' by human standards.'

Is life really that brutal? Is it really as simple as living and dying, often in ways that we might consider totally horrible? Often with oodles of suffering involved? Often with great pain and great waste of potential? Must this rational zoological outlook on life and death be quite so bleak?

It certainly sounded very bleak when I thought about it. But maybe I was looking at it wrongly?

A few weeks after corresponding with Anne it was toad season. The spring toad migration is one of my favourite times of the year. I generally like any animal you can pick up, intensely eyeball and probe, and put down without it being in the least bit bothered and, for me, this totally encapsulates what toads are like. With its dry, almost scaly skin, orange eyes and a slow waddling crawl, it has almost been designed via natural selection to fit perfectly into the coat pocket of an 11-year-old boy or girl. Toads are such resilient little things. Tough-skinned, rugged-stanced. No wonder they are rooted so deeply within the fossil record.

But toads are not so resilient that they can withstand getting hit by a modern human invention like, say, a car. If they get hit by a car they are not resilient at all. They simply die. Or they twitch their limbs in assumed agony for a little while and then die. Many of Britain's roads are littered with their corpses each spring. The TV quiz show *QI* (and I have no idea where they got this fact) says that 20 tonnes of toads are killed on Britain's roads each year, which is nothing if not incredible. But it's no surprise to me. There are a lot of European common toads out there and you could say that evolutionarily speaking the toads bet on the wrong horse: they went for poisonous skin and stamina over the speed and wariness that frogs, generally, possess. In a world of human vehicles they chose … well … 20 tonnes? That says it all.

Rescuing toads from roads is one of those activities that you'll normally find a friend of a friend does. Normally that friend of a friend is: a) eccentric; b) kind and loving, perhaps overly so; or c) having marital difficulties and needs an excuse to leave the house. Most toad patrollers are any combination of these three things. I am definitely one of these things (you can decide which, but I'm pleased to report that it isn't c). In temperate climates, on the whole, amphibian migrations take place during runs of consecutive warm, wet nights (particularly after or during rain), often in early spring. To see them yourself at this time of year, look at a Google map and pick out local reservoirs, lakes and big ponds. After dark, travel slowly and safely on roads near these freshwater spots and you will probably see them trundling along, particularly on more humid nights.

Toads are more picky about their breeding ponds than frogs. Whereas in northern Europe frogs prefer shallower ponds, toads appear to prefer bigger, deeper bodies of water. But such breeding spots are rarer than small ponds and this is a further source of bad news to toads: they must travel further to get where they need to go than frogs, navigating more and more obstacles, like roads and housing estates, in the process. Some populations of toads have been doing fine, it seems. Others – where roads are busier, for instance – are less fine. In Britain, with more and more roads, our toad populations are facing death by a thousand cuts; they're declining so slightly in so many places for so many reasons that barely anyone has noticed or is able to do much about it.

Each year during toad season there are five ponds that I normally visit, keeping tabs on how local populations are doing. Why do I do this? I'm not sure. It's partly out of duty, but also because I like to see and pick up toads and this is the only time of year I can really do it. It's exhilarating in some ways, driving down small roads, looking for tell-tale shapes like dead leaves that move slowly in front of my beams. It's like a really, really, really, really tame version of a night-time safari, which is to say it's still very exciting. Anne Hilborn would approve, I think.

I pull up to my first site. It's a small B-road near Great Brington on the western edge of Northampton. The toads here like to breed in the moat-like pond that surrounds Princess Diana's burial site, which is quite a nice thought (I think it's what she would have wanted). In many ways the site is typical: on one side of the road is a hill upon which a woodland sits in the distance, and down there, on the other side of the road and over a large brick wall, is where the large pond lies. Toads wake up from their winter slumber in the woodland and, *en masse*, make a move down to the water, crossing fields, hedgerows, a small ditch, and now this road. I think they must find little holes in the wall to squeeze through, but I have never actually seen them do this.

There are plenty of toads about tonight. Fifteen are already dead on the road. A large female has had her head squashed by a car and her unfertilised spawn has fired out of her rear end. This is particularly sad because female toads take more than three years to mature – as a result, to a meta-population, the life of each female really does count because it is capable of restocking tadpole numbers with such vigour. But not her. And not here. Most years I don't think much about all the death – they are casualties, and I'm here to try and help the living toads – but this year things feel slightly different. This year, the dead ones are as interesting to me as the live ones. I shine my torch on each one, assessing their size, their sex and their missed potential. My interest in death is changing me.

Toad crossings are always much sparser than you might imagine them to be. The toads move so slowly, they look more like an army of the undead crossing a graveyard than a sweaty tangled sexed-up throng like they appear on TV documentaries. But it is the number of them that keep coming that makes it all so impressive. For hours and hours, night after night, they keep coming from that woodland, heading over the road to the pond. By helping so many cross, by counting up the living and the dead, one gets a real feel for the statistical likelihood of survival, and, almost, the whole energetics of toad populations and what they bring to an ecosystem. You get a really good feeling that each life and each death matters for something. And also one gets a feeling for what death brings, in particular.

I think again of Anne and her super-rational view; the inevitability of suffering and death and how perniciously it appears throughout nature. Is this just the way nature has to be? Are animals doomed to suffer within never-ending food chains? Partly. But not totally.

Charles Elton is perhaps the scientist who has been most instrumental in our understanding of the ecological principles surrounding food chains, ecological niches and the concept of pyramids of numbers within ecosystems. Yet how he got such insight is interesting. It didn't just come to him. In 1921, early in his academic career, Elton decided to undertake fieldwork in Spitsbergen, an Arctic island with vegetation so low that Elton could tootle about watching predator and prey interact with little fear of humans. Elton spent a great deal of time watching the Arctic foxes go about their business, being the most easy animals to pursue and observe from a distance. And it was their movements that were to give Elton the insight he needed; insights that would later bed in whole generations of ecologists.

He watched and took notes. The Arctic foxes spent much of their time feeding on the birds – ptarmigan, sandpipers and buntings among them. So Elton observed them. He noticed that the birds fed upon smaller things: insects, grubs and seeds. In our busy world, it is easy to become drowned in the complexity of ecosystems and food webs, but there, on the tundra of that island in the Arctic, it was plain to Elton because there was so little else going on. It slapped him in the face. It was really, really simple: there were hundreds of thousands of insects, feeding thousands of birds, feeding hundreds of foxes. The observation that food chains organise themselves into layers like this was probably known to other scientists, and that something like a tenfold multiplier is often apparent between each layer, but until Elton came along no one had thought to ask … why? Why is life like this? Why does it form, on the whole, such predictable layers and food chains? No one had thought to properly ask *why* and seek to answer the question methodologically until that point; until Elton pondered that exact question in the middle of the Arctic watching foxes chasing birds and birds chasing insects.

There are other questions that come up when considering things such as food chains and food webs. For instance, why are large wild animals on the whole rare, but creatures lower down on food chains more common? Why could I see thousands of toads on a spring night yet Anne will only see a handful of cheetahs in a week or so, for instance? Famously, Elton saw in his mind's eye a pyramid of numbers to describe this. And indeed, for each 'trophic' level, there was often 10 times the number of the next step up: tens of thousands of insects, thousands of birds, hundreds of foxes. Elton was tempted to view this arrangement of trophic levels simply in terms of biomass so that, for instance, if you put hundreds of thousands of insects into a blender you'd have the same amount (biomass) of protein-shake as you would putting thousands of birds into a blender (please don't), or hundreds of foxes (stop).

But Elton realised something, something that modern biology textbooks remind us of frequently: this *isn't* what we see in nature. The volume of flesh in each trophic level isn't the same. The volume of flesh available decreases as one moves up the food chain. In essence, top predators take up less *flesh* than you'd think, by looking at the amount of flesh fleshing around in the trophic level before. It was the same with the mid-trophic levels. And the levels further down. In food chains, flesh seems to go missing … but where? Where does the missing flesh go? It was a mystery. Elton just couldn't work it out so instead he put the question out there in 1927 for other academics to solve. It took two decades before an answer was proposed.

The answer to the problem of the missing flesh came from two Yale scientists, Raymond Lindeman and Evelyn Hutchinson, who solved it by thinking of animals not as animals, but simply as calories. A parrot. A mayfly. A gibbon. A killer whale. Lindeman and Hutchinson viewed these creatures simply as spontaneous and temporary bags of calories. They viewed the life actions and behaviours of these creatures as being funded by the calories they have in their respective bodies, and they appreciated that these unusual-looking bags of calories could invest their calories in different things should they wish: finding more calories, investing calories in reproduction, or perhaps saving calories by sleeping. The point they realised is that you can't make calories out of nothing and that all calories come from something. All of us are going about our business frittering spent energy (often in the form of heat) from the burning of calories. Lindeman and Hutchinson understood that every hot breath that radiated heat from every animal that ever lived removed calories from a food web, starving the trophic level above. Starving the predators of more calories, in other words. I like considering this view of life, every now and then. In fact, I rather like the thought that every time I breathe out or walk around I am restricting the amount of energy that can be used by the predators above me in the food chain (fuck you, polar bears!).

Elton's early-twentieth-century ideas about ecology percolated, aided by Lindeman and Hutchinson, across the whole of biology during the fifties and sixties. Just as DNA had sucked biology into the realm of the chemists, Elton's ideas were among the first to pull biology into the world of physics, made more graspable by Schrödinger and his book *What is Life?* described earlier in this book. Yet, we rarely seem to talk so publicly about Elton and Lindeman and Hutchinson as we do Schrödinger. They deserve more of our celebration, I think. Without them our understanding of life, death, the universe and … everything … would be severely impaired. They were the first to understand that the laws of thermodynamics play out each day, like a grand marble run, in the ecosystems that sustain life. The laws of physics only allow for so many toads and so many cheetahs, it seems. According to this view of life, every suffering juvenile cheetah, left hungry and wanting, is a victim of universal circumstance. But within each, of course, are calories for something else. A lion or a hyena. A crow or a magpie. A vulture or a red kite. Eltonian physics led to a view that life is a convoluted opportunity for something else, constrained in its potential by our universal propensity for giving off heat. Should that make us feel a bit less morose about all the suffering? Maybe. But there is a hidden part of this picture. A missing part of the puzzle that is causing many modern-day ecologists to rethink Elton's big idea. For there is another place that all that flesh goes. It doesn't all flood out into the universe as heat. Some of that missing flesh can be found in the parasites.

For decades, parasites were the dark matter of food web energetics. Overlooked. Forgotten about. Downplayed. Yet now, many scientists argue that parasitism is actually evolution's favoured method of predation; taking from the living to give to the living, in other words. A number of studies support this once outlandish claim. In estuarine systems, for instance, the yearly productivity of flukes (parasitic trematode worms) is actually higher than the biomass of birds, according to evolutionary ecologists Daniel L. Preston and Pieter T. J. Johnson in their *Nature* paper, 'The evolutionary consequences of parasitism'. Then there is the fact that the biomass of the fungi that attack plants (in experimental grass plants set up in Minnesota) was comparable to the herbivores that graze them. And then there are the seabird islands. Some islands in the Gulf of California are two or three orders of magnitude richer than others in the populations of lizards, scorpions and spiders that inhabit them. The reason? The more seabirds an island supports, the more seabird ectoparasites they bring: mites, ticks, lice, fleas.

Preston and Johnson's verdict about the role of parasites in food webs is powerful: 'the classical Eltonian pyramid …

may need to be revised,' they write. It may be that animals in the middle trophic levels are the richest with respect to their parasite diversity, given that they offer places to larger communities of parasites (toads, for instance, have plenty of nooks and crannies for species to evolve into) and are susceptible to a greater number of potential predators (some of which the parasites will consider final hosts). It may turn out that, in nature, unlike in human life, everyone gets a piece of the middleman.

Anne Hilborn's super-rational view of animal death had stuck in my head. At first glance it had felt quite saddening to read her thoughts, but I realised that, this toad season, I was developing another way of looking at the sadness and the suffering. There is very little wasted potential when wild animals die, since life was being created all along. The toads had created and provided calories for generations of parasitic flatworms, mites, nematode worms and countless other creatures, plus many single-celled life forms yet to be described. And many of these organisms had been eaten by something else, powering the food web in a different direction. Their lives weren't 'wasted', in other words. In nature, very little is. And neither will those toads' lives be wasted in death, as the foxes and magpies each morning attest, as they scrape them off the roads before the flies get to them.

According to Froglife, the NGO that coordinates toad crossings (as they are called) in the UK, 76,710 toads were rescued last year (2014) by patrollers like me and 8,729 were observed to have already died that year, squashed on the roads by passing cars. In the nights that followed that first toad patrol this year, I counted 62 dead overall on the roads next to my sites. But by the end of that week it didn't seem so sad and miserable somehow. The toads weren't wasted because they didn't make it to the breeding pond. They had provided life all along. They were bags of calories, burning bright. They had given that light to others. And they would continue to burn within the bodies of countless other things, with a brightness ever so slightly dimmed.

Oh shit. I realised at this point in my journey that I was definitely starting to sound a bit weird and possibly slightly insane. The conversation with Anne hadn't really helped. The professional help wasn't seeming to do the trick. There was no doubt about it. I was turning weird. I was going somewhere with all of this, but where?

PART THREE

JOURNEY TO THE END OF THE SHITATITE

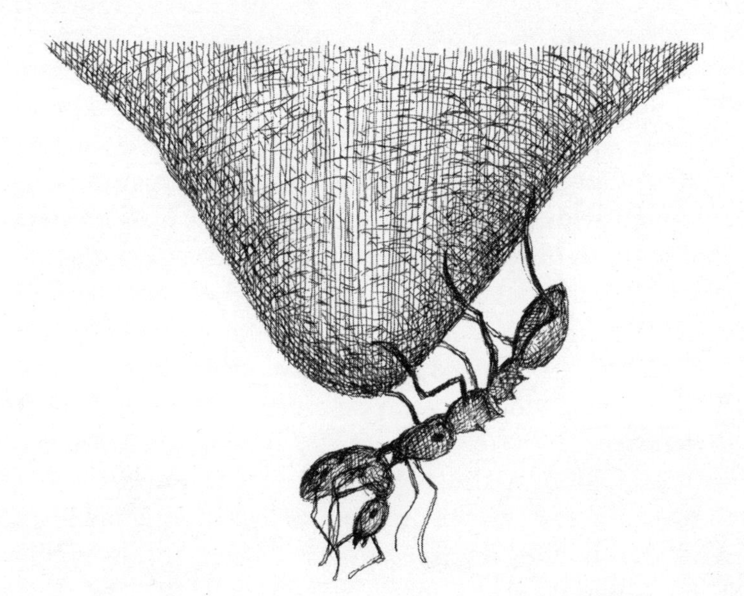

CHAPTER FOURTEEN

Bring out your Dead Ants

It's a laboratory that holds space for 80 or so students. Today, however, there is only a single human present. He sits in the corner alone at a bench on the far side of the room, hunched over something. He is studying it. The spring sunshine floods over him, coming through large, expansive windows. Students have picnics on the lawn outside; they chat and have fun. The birds sing. He notices none of these things. His mind is on other matters. We approach him. In his white lab coat he looks almost angelic – no, he looks like a deity. And to the ants whose lives he manipulates, that's kind of what he is. Adam leads me toward him and, as we cross the room, I attempt to work out what the hell he's doing.

'This is Stace Fairhurst,' Adam says as we get close to him. Stace gives me a big smile. 'Hi,' we both say. I go to shake his hand but at the last minute I realise he is holding a pair of tweezers in which a tiny ant struggles. Stace continues with his work. He carefully leans over his desk and, with his tweezers, places the single ant in a plastic tub on the desk in front of him. Inside the tub is a tube; a single corridor which splits into two, forming a Y-shape that looks a little like female reproductive anatomy. Stace and Adam explain what is going on. The experiment Stace is undertaking is simple: each worker ant is plonked at the bottom section of the Y and can move toward the left tunnel of the Y (which is being pumped with smells of dead ants and fungal waste) or the right-hand tunnel of the Y (in which there is nothing being pumped in but clean filtered air). Stace tells us that he has been placing worker ants in this contraption for days to see what they do and he is nowhere near stopping yet.

'Wait …' says Adam suddenly. We all stare at the ant in the plastic Y. It dillies and dallies at the entrance of both tunnels. It waves its antennae slightly. 'Wait, it's going to go …' Adam commentates. 'She's going to go … go … going … THERE!' The ant crosses a tiny line on the floor of the Perspex box, indicating that it has made its choice. This little ant chooses life: it avoids the corridor that smelt a little like death. Stace makes a note in his notebook and reaches for the tweezers again. The ant is pulled out of the box and is put back into the nest behind him. Another ant is taken out with the tweezers and takes its place in the Y-shaped contraption. This is just one of many experiments overseen by Adam and Stace (of the University of Gloucestershire) as they attempt to answer important questions about life, death and ants.

I should say early on that I already know Adam. I know Adam because he is Professor Adam Hart from the University of Gloucestershire, but also because he's often on the TV. This is the first time we've met in real life, though.

In real life it is even more apparent that Adam is the most youthful professor I've ever seen. He is enthusiastic about everything (even car-parking arrangements) and incredibly media-friendly. Throughout our interview he delivers complicated sentences slowly and arranged correctly, making them really easy for transcribing, which is a kind thing for him to do. Whenever I talk he looks at me with his full attention and does noddies like a pro. He's warm and open. Smiley. Funny. Knowledgeable and very lovely. Adam has invited me down to talk about how ants deal with dead colony members. In his lab he keeps many colonies of leaf-cutting ants, which are one of his many research interests. He sees a lot of death.

We stand there, all of us, watching the next ant as it goes into the Y-shaped box. This little worker ant tentatively moves toward the corridor that smells of death. Then it stops. It walks up and down a bit before committing fully and retreats back to the mouth of the Y. We wait a minute or so for it to do something. It does nothing. And then, quite confidently, it turns to the right and heads down the corridor in which clean air is being pumped. Another ant has chosen life. Stace scribbles the result down. I try to look at his sheet. 'Which corridor is winning overall?' I ask. Neither Adam nor Stace bite. Stace gives me a little knowing nod and a smile. 'We should really wait for the results, don't you think?' he says. 'Of course,' I say quickly. 'Of course.'

Adam beckons me over to one of the ants' nests on the table behind Stace. The nest is wonderfully contained inside a long glass tank that is about 150cm by 30cm. The colony itself sits on a wooden platform in the middle of the tank, as if on an oil rig. It is held on four stilts above about two inches of water; an island from which the ants cannot escape. Upon the platform thousands upon thousands of leaf-cutting ants mill about on a pile of privet leaves. They nibble these leaves into small chunks which they carry back to the nest, which is housed within a series of plastic tubs that look like spent Ferrero Rocher boxes. In these tubs the leaves are being composted, their fungal food is growing and the next generation of ant larvae are being raised and fed. There are many, many ants in the tank and most of them look like they have no clue what on earth they are doing.

I peel my eyes away from the throng scurrying around on the pile of leaves and scan beneath the platform, looking at the water below. Adam tells me that there is a tiny bit of detergent in the water so that ants are unable to travel upon the meniscus and make an escape. My gaze returns to the ants on the top of the platform. Some of the ants stand on its edge and appear to look wistfully up at us through the open top of the glass tank, waving their antennae slowly back and forth as if to channel their thoughts of escape into the universe. There look to be tens of thousands of ants in there. 'Have there ever been escapees?' I ask Adam tentatively. He tells me a story about leaf-cutting ants climbing up the filter pumps in someone else's lab, and how the staff came in the next morning to find a line of ants carrying tiny chunks of toilet paper back to the nest from the restroom. I don't know if he's joking or not. I let out a nervous little chuckle and quietly check my shoulders, my hair, my face, my legs, my torso and my upper arms for rogue leaf-cutting ants.

There are many tanks of leaf-cutting ants used as part of Adam's research. Each is a city, a self-contained universe for ants. Each is a sample. 'Each queen in every one of these tanks may live 20 years or so,' says Adam as we look at a different tank. 'Given time they can turn something small into something enormous.' We peer into another tank. 'That's a hundredth of the size of a real nest,' he says, smiling again. 'In real life these nests are just … they're just enormous! They're absolutely massive!' Adam is loving this, and his passion quickly worms its way into me. We both talk in excited tones, like we're still at school. 'They're the size of a *house* underground!' he says. 'A hundred chambers, each the size of a basketball, each stuffed with their special fungus …'

He talks fast and full of excitement and, for a few moments, I decide to just sit and listen to him, like I'm watching him on BBC2 or something. 'And you've got massive amounts of waste in there,' he says, pointing at another colony. 'Massive amounts of waste coming from out of the nest. Massive amounts! Just consider the amount of dead ants that come out of these nests!' He takes a breath. 'The Queen? Sure, she might live 20 years or so, but the workers? The workers …' He smiles again. 'They might only live a few *months.* And, listen to this …' He looks at me quite eagerly again. 'In a wild nest there might be *eight million* workers, and each of those eight million workers might only live for a few months … Imagine that!' I imagine a long train of dead ants being carried out of the back end of the nest. 'So the death rate is just … well, it's *huge*,' Adam concludes.

All of the ants in Adam's tank are from one species of South American leaf-cutting ant. Adam tells me that in the wild these leaf-cutting ants excavate waste chambers beneath their nests into which their waste is then thrown. Leaf-cutting ants are renowned for this strange behaviour; other ants don't do it this way. According to Stace and Adam, other ants create enormous waste piles outside, usually downhill and often near water so that their chemical signatures can be washed away. This is where the dead ants are carried, and these waste mounds are often worked by the oldest, dying, workers. Apparently, sometimes when ants have finished up burying their own, they quietly dig their own graves and expire into the mass. 'You see this sometimes even in labs,' Adam tells me. 'The oldest workers dig deep pits into the waste as they're working it, and then look back up and think "*Oh wow, this is comfortable*" and basically die sitting in their own graves.' We laugh at this. I guess it's because we consider ants so unthinking – so resolutely without aim or purpose – that it's hilariously predictable that they would be so laissez-faire about their own death, making a polite space within the rubbish dump in which to die. Ants seem to consider death like a robot might.

Stace reaches for another ant and we gather around once more to see what the next little worker ant will do. 'I think there is one thing that strikes me with all of this talk about death,' Adam says as we watch. 'When we look at death in nature, and even with death in humans, so much of the ceremony around death is fundamentally about the single fact that a dead body is a big stinking problem that needs to be solved immediately.' We all nod our heads. 'A stinking problem,' I hear myself agree. 'And that's what ants do so well ...' Adam smiles another of his little grins. 'They solve it. They solve the problem of death,' he says. 'I wouldn't say there's no *ceremony* with the ants, though – lots of ant species create dead piles that aren't unlike cemeteries in some ways – but basically they're just getting rid of bodies as quickly as possible. And that's it.'

We stare at the ants for a bit longer and a queue of questions forms in my mind. There is one big question I want to ask first but I'm waiting for the right moment. One that's top of the list – a question that betrays my own morbid fear of being buried alive. 'How do ants know that another ant is ... dead ... before they lift it up and cart it off to the dump?' I ask. I have visions of ants looking at their sleeping siblings toying with the idea of taking them 'downstairs' because they appear to be sleeping particularly deeply. Adam and Stace look at one another. 'That's actually a really interesting question,' says Adam. He takes a few moments to organise his thoughts. 'Hmm ... it depends. Whether it's because they identify something about another ant that suggests it is dead, or whether it's because they identify something about it that's *not alive* ... it depends. It could be one or the other. Or both.' The truth is that no one really knows for sure. Adam introduces me to the idea of what he calls jokingly (and some scientists take rather more seriously) the notion of a 'vital smell'. 'There's a lot of different studies looking at that exact question of whether it's the lack of a smell of a living ant – the vital smell – that's key,' he explains. 'But there's necromones too, which are the smells that are distinctive to death. Oleic acid is the classic. If you soak a rice grain in oleic acid they'll treat it like a dead ant. They'll do this every time.'

But there are other signals that ants give to one another to prove that they are alive and well. As we stand and watch the ants running up and down the platform it's obvious to see that they are constantly touching one another with their antennae (Adam calls this 'antennating' – a word I hadn't heard of until then), feeling for behavioural responses and cues and signals that only living ants give off. 'I suppose we could anaesthetise some ants and see what happens to them to stop them antennating, or you could wash the smells off them, and see what the other ants do.' He thinks for a moment. 'It's interesting ...' He trails off in thought and I know he is mentally imagining a paper he will one day write.

I have immediately taken to Adam, and the pondering moments like this one are why. Suddenly his train of thought jolts onto a new groove in his stream of scientific consciousness. 'There is a bigger problem with death,' he says, 'perhaps more problematic than how you get rid of the body.' He beckons me over to another tank. 'When an ants dies you've also lost a worker – there appears a sudden gap in the workforce which needs to be filled. How they fill this gap is fascinating.' Adam explains that too many leaves coming into the nest sees the workforce shift toward nest-builders and not enough leaves coming in means the ants shift the workforce toward creating more foragers. 'Death is a big part of their life,' Adam says thoughtfully. 'Death is a big part of how they organise themselves.'

We go back to Stace, who has just watched yet another ant make a tiny choice. He leans over and gently places the ant with his tweezers back onto the wooden platform in the tank behind him. I try and keep my eye on where this particular ants goes now it is back in the nest, but it is impossible. It has become an indeterminate part of the great throng of workers swarming all over the place. Just another ant.

Adam calls me into the next room. This one is a smaller lab. On the far wall are six glass tanks on tables organised into a long row. These tanks look bigger. The colonies are arranged in the same way: densely populated platforms, nests in Perspex boxes – the sea of death below each one, keeping them all from flooding out into the labs and into human lives. In these bigger tanks we can see more easily how the ants dispose of their dead, since each of the tanks is due for a clean. In the waters below each colony are the submerged ghostly corpses of ants that have been tossed into oblivion over recent days from the platform above. 'They're the dead that have been jettisoned off the side; dropped by workers from off the bottom of the colony,' Adam says sombrely. The dead ants lie motionless on the bottom. I get nice and close to one or two, looking through the glass at them. Their legs are curled up around their bodies, their head arched back and their jaws still open. They wear expressions of suffering that I'm well aware that they do not have (being that they are ants and are limited in the number of facial expressions they can have, namely one or possibly two).

I look up at the underside of the platform on which the colonies sit. Hanging off each wooden platform are three or four little brown stalactites. They appear to be made of tiny grains of stony mud. I look at them really closely. A handful of ants crawl up and down these strange stalactites, and I notice that one or two carry little blobs of grit, which they drop into the water before climbing upwards and back to the colony. Adam sees me staring. These mud formations that dangle off the bottom of the platform are what Adam calls 'shitatites' because they are, in fact, made up of congealed faeces and dead fungus dropped from the platform above. Over weeks and months occasional grains of these droppings have coalesced and become solidified – they now form a drooping mass of hard faeces and fungal waste that has become a handy new pier upon which new ants can clamber down to get nearer to the water below if they so wish.

We spend a few minutes in silence, watching the shitatites. I watch the journey of a single worker ant: it walks from the rabble of ants walking around on the top edge of the platform and makes its way over the edge and round and underneath. It walks effortlessly upside down, almost seeming to relish the quiet and open space, away from the din of its workmates above. It makes a comedy path around on the lower edge for a few moments, going left, going right, then retracing its steps. Stopping. Spinning around. Going back. It really doesn't seem to have much of a clue. It finds a shitatite. It climbs down the shitatite until it is only a few centimetres from the murky waters below, and for a few moments it hangs upside down off the tip, held by only three of its legs. In that moment I find myself hoping it will fling itself off into the water, which is a strange thought to have because I'm normally quite kind to animals, but I like the idea of telling a story where a worker ant shows signs of suicide. A second or two passes. It does nothing. Still nothing. And then, something: the ant pulls itself back onto six legs (still upside down) and opens up its jaws. A tiny bit of grit falls into the water below, where it slowly sinks through the water and rests on the bottom, like a heavy stone thrown into a lake.

The ant now attempts to ascend the shitatite back to the colony above. It has trouble doing this; there are a few moments where it desperately struggles to find a foothold on its climb back up. Its first pair of legs fail to find a grip and it hangs there for a few moments on three legs again. The tiny ant looks momentarily like Luke Skywalker hanging off Cloud City at the end of *The Empire Strikes Back*, but it finally resolves its situation; it finds some sort of inner strength and hauls itself back up the shitatite to join the rest of the colony on the platform above. Within two or three seconds, amongst the myriad ants swarming over the leaves, we are unable to see where it goes. It has become them again.

Adam and I watch the great swirling mass of workers for a few more moments. Then a few more. I realise we haven't spoken for at least a few minutes; we have become drawn into their world. It's me that eventually breaks the silence. Almost dreamily I ask: 'Do you ever get bored of ants, Adam?' Adam speaks equally dreamily in his response. 'No, they just seem so …' He really thinks about his answer to this question of whether ants are boring. He really contemplates it. 'No,' he says. 'No, I would never find them boring. There's a determination to what they do. Just look at them …' He beckons to me to come closer to the tank in front of which he stands. 'Come up close to this one,' he says. I come closer. 'Look at a single ant. Pick out an ant with your eyes and watch it.' I pick out an ant and watch it. 'Most of the time what they appear to be doing is either very little or … they're just *irrational.* Watch this one with the leaf …' He points to a single worker near the edge of the platform holding a flag-shaped piece of leaf in its jaws. It stumbles this way and that with the weight of it. 'Look at it,' says Adam. 'It's going the wrong way for starters. It's heading away from the colony, coming precariously close to the edge of the platform. It doesn't really know what it's doing,' he chuckles. He's right. It looks totally lost. 'And look at the others nearby. They're just milling around really. They don't know what they're doing either. But then step back.' We pull our heads back and look at the colony as a whole. 'Look at it as a colony,' he says, 'and you can see, clearly, on the whole, that there are patterns: a line of ants carrying leaves into the nest.' Indeed, there is a pattern. Leaves going in. Ants coming out. 'Stepping back, you can see it,' he says.

He's right. I had viewed the ants like clockwork soldiers in an army, regimented and unfaltering, but really they reminded me of shoppers in a mall – all doing various things and visiting various shops but overall spending money and helping the mall pay the bills. It was complex and very messy. But there were patterns. Leaves (on the whole) move toward the colony. Soldier ants (on the whole) stand guard. Worker ants (on the whole) head toward the shitatites and drop little balls of faeces and their dead into the void. On the whole, the colony works. It is mesmerising to see them like this in these tanks, in a world so observable to us humans. They'd make great pets, I think, if you have ample supply of fresh leaves and possess understanding friends and family and pets and landlords and have no children and possess good access to detergent. Great pets.

'A dead body is a big stinking problem that needs to be solved immediately,' Adam had said earlier, referring to the situation that many colonial animals find themselves in when it comes to death. The solutions that natural selection has come up with to solve this problem are surprising in their diversity. In many bee species, for instance, there are 'undertaker bees' who are specialised in the role of removing the dead (and there apparently seems to be a genetic component to this behaviour, too). Social wasps also display such behaviours. And then there are the termites. In recent years one research team has exposed a great deal about the unusual behaviour of termites when it comes to dealing with dead nest-mates. Their study was rather simple, and not totally unlike those that Stace and Adam were undertaking at the University of Gloucestershire. A research team from Universiti Sains Malaysia and Kyoto University investigated the responses of four different species of termite when introduced to termite carcasses in various states of decay. Traditionally, the scientific literature had it that termites avoid their dead (called, predictably, necrophobia), but their research, published in 2012, uncovered the much more dynamic world of complex death-management that occurs in termite mounds.

Upon artificially introducing termite corpses to a nest, the responses were often found to be the same: the first workers to come upon the carcass would immediately pull back and recruit other (unexposed) workers over for a second opinion of the corpse. This was standard behaviour. From this point forth, however, the four termite species showed different responses. Workers from two of the species reapproached the termite corpses and investigated them. If the dead termite was deemed to be recently killed, and decomposition was in its early stages, these corpses were carried off and 'recycled' (they were eaten). But if dead termites had been left for too long, the termites considered them inedible and they were carried off as waste.

The other two termite species behaved totally differently when coming across the dead. There was no consideration; no recycling or waste removal of dead termites. No, none of that. The other two termite species were observed to simply 'wall off' the locations in which the dead termites had been found, creating disused tunnels, sometimes filled with dead colony members, never again to be entered. Why do they do this? It could be that this behaviour reduces the likelihood of disease and parasites spreading through a colony. No one's quite sure yet.*

*I wondered about other social animals. How do they deal with their dead? I wondered about the naked mole rats that featured earlier in this book, for instance. Finding nothing in the literature about this I made contact with one of the world's leading experts in naked mole rats, Dr Chris Faulkes (a Reader in Evolutionary Ecology at Queen Mary, University of London).

'A big stinking problem that needs to be solved immediately' – hearing this line made me consider the human response to dealing with our dead. It was inevitable really that I would choose to bring this up. The elephant in the room, I guess. Though we are not social in the same sense as ants or termites, we are social in many of our habits and our housing. The same stinking problem applies equally to us as to them. Here are some figures which I have pulled from Bernd Heinrich's wonderful book, *Life Everlasting*, about humans and how many of us choose to deal with our dead in the modern Western world. These statistics are as follows. Of the 22,500 active cemeteries in the United States, the combined materials they use to cater for the human dead are: 30 million square feet of timber; 100,000 tons of steel; 1,600 tons of reinforced concrete; and (get this) almost one million gallons of embalming fluid. But that's fine, you're thinking; you were going for a cremation anyway, right? Well, annoyingly, cremations are hardly the eco-alternative one might wish for. According to Heinrich, the combined fuel used to burn all those bodies is enough to power perhaps 80 trips to the moon and back, not to mention the damage that the airborne mercury may cause in pollution (cremations are the second highest cause of airborne mercury in Europe).

It's quite depressing. Be under no illusion: in the developed world, death is big business. It's been monetised. It really has. The funeral industry is worth $20 billion to

Here is his response: 'Yes, an interesting question, and annoyingly one that we cannot answer! I can find no reference anywhere to what may happen to the dead animals from the limited fieldwork that is published (on naked mole rats). My colleague Nigel Bennett has never dug up a corpse in 30 years of fieldwork on various mole-rat species, and neither have I. It is tempting to speculate that they may drag corpses into a toilet chamber where they become buried and eventually sealed off when a new chamber is excavated. But nobody knows.'

the annual economic activity of the United States. Twenty billion dollars. They want you. And we want them, because we think our life is worth it. And because we want to get rid of the big stinking problem, I guess.

On the drive home from visiting Gloucestershire I thought long and hard about my own feelings of what I want done with my body after death. I decided that I had half a mind to let the ants have a go with me after I'm done. In fact, if I could distance myself from the disgust for long enough, I came to quite like the thought of giving my body to a colony of scavenging ants. Of my flesh becoming the flesh of eight million ants in my death. Me: part of the action. Part of something big. Part of something organised. For the first time in my life, I realised I could have purpose in death. And I could choose what. I think, if I really considered it, I could make some quite interesting life from my dead body; I just have to opt for what life I want to crawl out of me. It felt strangely life-affirming to think like this; not at all what I had expected when this journey into death had begun. My body: a vessel. Not a metaphor for anything. Not a figurative vessel of memories; not a vessel on its way to a higher place or a spiritual heaven. A proper vessel. A physical vessel. A vessel for something else to make use of after I've finished with it. I just had to make a choice. The question is, which corridor would I choose: the left or the right?

CHAPTER FIFTEEN

Mourning has Broken

This is the story of a chapter that could never be; and it begins at the end. I am sitting at a desk in a Best Western hotel looking at myself in the mirror while the cognitive scientist Dr Alex Thornton speaks to me on the phone. My shoulders are sagging. He's hammering me hard, and it hurts because I know he's right. In the mirror I look tired. Tired about whether, and to what degree, animals know death like we humans do. 'How do you *know* that my grief is the same as yours?' says Alex. 'How do you know that I *feel* things like you do?' This is the second time he's asked me this on his crackling line. 'How can you *tell*, really?' It dawns on me that Alex is enjoying this. Alex is chewing on me like a great big philosophical steak. My head hurts. I pull at my face in the mirror. There are bags under the bags under my eyes. Death is taking its toll on me. The notepad on the hotel table is dry. Alex goads me a little. 'Go on, Jules …' he says. 'How do you *know* that I feel love and mourn and grieve just like you do?' A pause. I want to choke out a little sentence but nothing much comes out. 'You don't, do you?' says Alex. A few more seconds pass. 'You don't know. You can't know,' he concludes.

Alex is challenging me about the popular notion that many animals grieve and mourn and feel loss and sadness like we do, and his view is that it's not something we may ever know. I had rung him in desperation. This has been by far the toughest part of the book to write. Six months of research into animal mourning has got me nowhere. Absolutely nowhere. I'm drowning in a sea of anecdotes that float like chunks of ice around me; as soon as I clamber onto one, I lose my footing and slide off back into the sea of uncertainty. I was desperate for something firm to hold onto. Good evidence. Any story about animal mourning with a sample size greater than one or two, for instance. 'I guess I can't,' I admitted to Alex. 'I guess I can't say with certainty that you and I feel the same things in the same ways.' 'Exactly,' he says.

I consider telling Alex about the time when I thought my four-year-old daughter totally understood the death of her great-grandmother but then later I'd caught her creeping past great-grandmother's old flat attempting not to wake her, even though we'd seen her being cremated 90 minutes beforehand. On reflection I choose not to mention it. At this late stage in my research I can recount hundreds of tales of animals that show hallmarks of mourning or grief but I honestly cannot say with any certainty whether, and to what degree, they may or may not be real and true understandings of death as you or I know it.

I drum my fingers on the desk, realising that this chapter may be bound for the bin. Alex is right: if I can't fully understand and guarantee his feelings and emotions, and he mine, then how can we know cognitively what animals think when those around them die. Alex has me philosophically pinned, like a metaphorical wrestler in a WWE ring. I look around for a metaphorical folding chair which I can swing at him violently, releasing me from his grip. 'What about chimps?' I say. 'They mourn … I've seen it on telly.' Even as I say it, though, the words sort of limp out of my mouth. I begin to recount to Alex about a haunting nature documentary I had recently watched in which a female chimp carried around her dead offspring for days and days on end, seemingly unable to accept its death. The tiny corpse had dried up and become waxy, almost mummified. It was heart-wrenching to watch. The story had formed a centrepiece of a BBC show crafted to make us feel closer to our primate cousins in terms of our ancestry and shared cognitive skills. Alex's response was instantaneous: 'Sure, it carried around its dead offspring … but a simpler explanation is that she didn't know that her offspring had died,' he says dryly. This leaves me momentarily speechless. What? I think. What Alex is saying seems just so *heartless*. Alex reads my silence. He carries on: 'Look, there's a massive tendency for us to ascribe these sort of things to creatures that we feel we're related to,' he says. 'But if we saw the exact same thing happening in a mouse most people wouldn't think twice about it.'

He pauses and then continues on the theme of chimps. 'Listen, there's been about six million years of evolution since we split from chimpanzees: plenty of time for novel traits to evolve. And there are all sorts of things that are different in humans and chimpanzees. Do I think that in this case, in mourning the dead, we're similar? I don't know. I don't think there's any way that we can really answer it.' He goes a little quiet for a moment, considering his words. 'One thing I would say is that people have been studying chimpanzees in the wild intensively since around the sixties when Jane Goodall set up her studies, and in that time there have been a handful of these supposed 'grieving' events, but chimpanzee researchers have come across loads of dead chimps. Chimps die,' he says bleakly. 'If apparent grieving events happen it's exceedingly rare. It's not like in human societies where almost every time someone dies there's massive social disruption.'

I agree with him about this. When humans die the behavioural responses that this death elicits do seem incredibly powerful and are easy and clear to observe: there is often wailing, screaming, crying, sobbing, holding, touching, embracing, for instance. Sometimes this can go on for days and weeks and months and years. We seem *different* to the rest of nature in this respect. But I struggle internally with Alex's super-rational approach to understanding animal grief. Something about it doesn't sit right. I've spent my life communicating to people that humans are animals; that, in nearly all ways, our bodies and brains work to nature's rules and that we have evolved (mostly) through natural processes that apply equally to mushrooms, midges or marmosets. Yet, here I was entertaining the idea that we might be the only creature on the planet to understand truly the finality of death. It feels heartless, I tell him, to suppose that chimps carrying around their dead chimp babies might not be feeling emotional pain like we do. Alex disagrees. 'I don't see it like that,' he says firmly. 'Why is it heartless to admit you can't answer something? For me, the excitement in studying nature is that there's so much we don't know. Sure, some questions are really hard to determine but they're fundamentally answerable. But some are fundamentally unanswerable.' He pauses. 'This is one of those fundamentally unanswerable questions,' he says firmly. 'To me, it's much more satisfying to be honest and say "*Look, we don't know what's going on here*" rather than saying "*OH! IT'S DEFINITELY MOURNING!*" because once you start down this line, you're no longer in the realm of science, you're in the realm of faith.'

Pow. And that was the knockout blow. The metaphorical bell rang: ding ding ding. Nothing hurts a scientist like the F-word. Nothing. After we hang up, his words continue to sting like a wet slap across my cheek. Science or faith? I think. I sit on the hotel bed looking at the pile of reference books I've been lugging around for what feels like months, dancing with an uneasy thought …

I had got the idea to visit Simba after talking to the famous bird-handlers Lloyd and Rose Buck. I had explained to them that I had once, in my car, accidently run over a jackdaw and that I'd parked up and watched its mate come from out of the bushes to inspect its dying partner. I'd always wondered what had been going through the surviving jackdaw's mind during this moment (in fact, I based a chapter on this particular bird in *Sex on Earth*). Was it feeling loss at losing its breeding partner? Did it understand the finality of the event that had just taken place? How much distress was it in? Or could it, perhaps, have simply been inspecting the eyeballs; weighing up the energetic value of the flesh that lay before it, maybe? Lloyd and Rose said they had something to show me. I drove down to Bristol to see them.

Simba was a crow, an 18-year-old carrion crow. As I stood in amazement, he perched on Lloyd's arm and looked at me and Rose purposefully. I had rarely been so close to such a stocky bird. His head was surprisingly furry; his waxy black feathers were neatly arranged down his wings and across his back like a leather cape. His black legs looked like they were covered in PVC, almost like he was wearing motorcycle trousers. He strode up and across Lloyd's arm and shoulders like a miniature Darth Vader marching up and down a starship bridge. It was intimidating being around Simba. It really was. I felt he might suddenly choke me with his mind.

'Rose, go and get the feathers …' Lloyd asked, smiling excitedly. Lloyd looked around at me. 'Watch this,' he said, his eyes widening. 'You'll like this …' He grinned. Rose walked into a nearby shed and pottered around a little behind the scenes. She came out and stood in front of us and then pulled a handful of crow feathers from out of her jacket theatrically. She threw the feathers on the floor in front of Simba. And then it happened. It was so sudden. A behavioural change came over Simba that I had never seen before. The atmosphere became suddenly charged. Simba paused, looked down at the feathers and immediately ruffled his feathers out. Deep from within his body came a series of measured long and deep caws. 'CAW CAW CAW CAW' Simba squawked. There was a rhythm to it. 'CAW CAW CAW.' The warm breath was visible from his throat. 'CAW CAW CAW.' With each exaltation Simba threw his head upwards, keeping one eye right on the mass of wet black feathers. It seemed to terrify him. 'CAW CAW CAW.' There was a rasping quality to it; it became a kind of roar. 'CAW CAW CAW.' Not what I'd expected from a bird at all. 'CAW CAW CAW.' It echoed across the pond in front of us, silencing the hordes of waterfowl that had honked and hollered constantly since we arrived. Upon seeing the feathers, it was as though Simba had momentarily become possessed. It was … a response. A behavioural response.

'He never reacts this way to anything else. Only black feathers,' Lloyd shouted over the din. I barely heard him. 'He never makes that call for any other reason,'

Rose said. Rose picked up the feathers from the floor, returning them to a pocket in her jacket. Simba immediately stopped making the noise; there was a moment and then he went back to normal. He carried on walking up and down Lloyd's arm as if the whole thing had never happened. Rose and Lloyd looked at me. I had an idea. I picked up a nearby clump of white downy goose feathers and threw them down in front of Simba. Silence. Simba looked at them briefly, then continued striding around busily upon Lloyd shoulders. Then Rose chucked the black feathers on the floor once more and we all watched as Simba went bananas again. Rose picked up the feathers again. Simba reverted to normal as if nothing had ever happened. It was incredibly interesting to see so clear a response. What it meant was … anyone's guess. Was this reaction typical? I thought. Was this behaviour common to crows and other corvids? What did it say about the corvid state of mind? Was it related to death, somehow? I had questions. More questions.

Throughout my death-journey, the question I have been asked most by friends and family is: 'Are you going to write about when animals mourn?' In recent months people have become really drawn to telling me their anecdotes about when their pets mourned or grieved after the loss of family or their kennel-mates. People have told me about horses they knew and how they buried their dead stable-mates in hay. They have recounted stories to me of dogs that stopped eating at the death of their owners, and stories of cats losing their hair after their kittens were taken away. People have sent me links to videos of swans committing suicide after the death of their partner; police dogs resting their paws on the coffins of their dead police officer owners. Elephants, horses, dolphins, cats, dogs: all behaving strangely around corpses of their own kind or even those of other species with whom they might have bonded. Rabbits that pine when split up. Dogs that apparently become anxious and restless at the same time that the house-cat is being put down. 'How can it not be sadness?' they say. 'How can it not be loss like we feel loss?'

I couldn't answer these questions. It really is a difficult phenomenon to confront scientifically. I realised this early on. I'm sure there was something going on with their anecdotes – in fact, I had agreed with my friends and family that there was probably something afoot. But what? Their stories were anecdotes and I wanted more than that. I wanted some science, I guess. But it was proving hard to find. Their anecdotes told me that some animals *can* and *do* show signs of apparent distress caused by, or linked somehow to, the death of conspecifics or offspring or owners. But their anecdotes couldn't tell me why it happens, the mechanisms through which it happens, or how common or easily predictable such behaviours might be. Dogs don't always show distress. Neither do cats or horses or monkeys. The whole issue of animal grief was problematic for scientists, as Alex warned me later on the phone. But Simba had come along. What I had seen in Simba seemed a simple test: a stimulus response that seemed repeatable, predictable; that might tell us something about the corvid understanding of death or its association with it. So Simba kept my interest for a while.

I had assumed that the kind of raised response to feathers I had seen in Simba might be very well known in corvids but … it wasn't. It didn't seem to be, anyway. I spoke to friends and colleagues who knew nothing about it, who had never seen this behaviour for themselves and had never heard about it from others. As I searched I could find only two references to this strange behaviour in the literature. The first is this: in Marzluff and Angell's *Gifts of the Crow* (a wonderful account of the intelligence displayed by corvids) there is a story about a Native American dancer leading workshops for tourists at an amusement park. When using black dyed turkey feathers to make headdresses, he is said to have heard a wild crow cawing from afar, and then watched as a mob of angry crows arrived. The birds mobbed the public, scolding anyone who wore black feathers in their headdresses. Only when the gentleman leading the workshop apparently decided to call an end to the event did the corvid bombardment cease. Apparently he never used dyed feathers in these displays again.

The other mention of corvids becoming agitated around black feathers comes from the greatest ethologist of his (perhaps all) time, Konrad Lorenz. In *King Solomon's Ring* he recounts a similar distress response from jackdaws after taking his wet black swimming trunks out of his pocket and giving them a shake after having a dip. He was the subject of furious attack for doing this; a reaction he put down to his floppy wet swimming trunks looking a little like a floppy dead jackdaw. So intense was the aggression in this encounter that the jackdaws drew blood from Lorenz.

And that was it: two cases – a Native American dancer and Konrad Lorenz. These were the only two cases I could find of this strange corvid response. I spoke to other corvid experts and none had seen this behaviour with their own eyes. No one could give me much to go on. Simba was set to become another anecdote of hundreds where animals behave in surprising and strange ways upon being shown things that remind them (or us) of death.

There really are hundreds of anecdotes out there. Log in to YouTube. See for yourself. There are a host of strange and obscure videos of animals responding in unusual ways when exposed to their own mortality or the mortality of those with whom they have bonded. Amongst my own personal gems is footage of a 'hero dog' rescuing an injured dog from a busy Chilean highway. It drags the other dog between busy lanes, beeping cars and screeching traffic, to safety on the verge. There's another amazing video of a macaque rescuing (and seeming to resuscitate) a fallen troop member that has been electrocuted on train tracks. The macaque picks up the steaming, apparently dead, body in its arms, pulls it about a bit, pushes it, pokes its face, squeezes its torso then chucks it in a puddle a few times, and, hey presto, the electrocuted dead monkey comes back to life (though it looks actually very ill, even after being revived).

There are other videos like this. There's one haunting video in which a Hawaiian green sea turtle lumbers out of the sea to 'pay its respects' at a little grave made by locals for another turtle (a female they had named Honey Girl) which had been found dead on the beach in the days previous. The turtle appears to struggle 10 metres or so out of the water and rest its head next to the grave, gazing up at the photo of Honey Girl on the memorial ('It was almost as if he was coming up to say goodbye,' says a local resident in the news report). These videos have thousands of hits, but while watching each one I found myself wondering what the fate is of videos in which animals don't do much, or fail to display the behaviours we class as mourning. Presumably no one wants to watch videos like that.

I continued searching for an answer to explain why Simba had behaved so strangely upon seeing the black feathers. There had to be something in the literature to confirm or deny it. There had to be. And then … I found something. And it was in the scientific literature, too. This research was on another corvid, the western scrub jay – a close cousin of the carrion crow. The western scrub jay is a native of western North America, a perky medium-sized bird with bright blue head and white throat and a 'harsh and scratchy' note to its voice. Like the carrion crow and other European corvids, the scrub jay is a common resident of suburban and wooded areas, in which it hunts in pairs or family groups in spring and summer and within larger non-family groups away from the breeding season. It does appear that scrub jays respond to death, and here's how it happens. The experiment (undertaken by experts at the University of California and described in the journal *Animal Behaviour* in 2012) placed a number of objects – including real and stuffed dead jays and stuffed specimens of great horned owls, along with some control objects – into different residential backyards. The researchers watched how each wild group of scrub jays reacted to each novel object. First, they carefully put a stuffed owl on the floor at one site (a natural predator). The scrub jays responded predictably, congregating together and occasionally swooping upon it in a mobbing behaviour. This was expected. Their tests showed that scrub jays also behaved this way upon being shown a stuffed scrub jay. This was expected too; scrub jays often treat outsiders with trepidation, ever-suspicious of potential lone guns and resource-stealing intruders.

But when the researchers placed a western scrub jay that was clearly dead in the middle of a colony they noticed something not expected at all. The scrub jays in the study population were observed to congregate around the dead body, forming cacophonous aggregations. Among the calls heard by the researcher were 'zeeps' and 'scolds' and 'zeep-scolds'. Apparently these information exchanges were so intense the animals were observed to cease foraging for food for hours at a time. They were rattled, it seemed, by the sight of a dead scrub jay. It made them react. It made them respond. And they did it again and again in a predictable way.

Evidence. Finally, some sort of evidence. Why do they do it? We don't know. What do they get from it? We don't know. The authors of the study suggest that this behaviour could alert group members, warning them to a potential danger (a kind of 'WHO KILLED THIS BIRD AND WILL THEY KILL AGAIN?' noise), but of course this is a hunch requiring further scientific scrutiny. I like this study, though. It seems so repeatable. Something any backyard owner could action (should they find themselves in possession of a stuffed owl or a dead scrub jay). But even this type of research is steeped in confusing language. For instance, the authors of the study use the popular parlance of 'funeral' to describe such aggregation behaviour in corvids, and this is the only point at which I start feeling a bit uncomfortable. A funeral? A funeral, really? The term, applied to animals, makes me feel … I don't know … uneasy. Who let that get through? We seem so desperate to make them like us that even our scientific language seems somehow to jump the shark.

If there is one thing that the scrub jays taught me, it's that if you want to obtain predictable, strong conclusions about how animals respond to death you need a good sample size and a neat experimental method. In most cases of apparent animal grief, this is sadly lacking. For many animals, that's fair enough – it may prove too difficult to test. In wild elephants (the most famous of animal mourners), only once has anything close to such a study taken place, to my knowledge. This study used bones.

In *Elephant Memories*, the famous elephant researcher Cynthia Moss recalls her initial surprise at seeing elephants respond to the bones of familiar family members. After bringing back to her camp the jawbone of an elephant matriarch she was amazed to observe (a few days later) the matriarch's family approach the camp and the matriarch's seven-year-old son pay special attention to the jawbone, probing and stroking it and turning it over gently with his feet. Moss was interested to test these observations experimentally so she set about, with colleagues, testing their responses by presenting different skulls of megafauna (including those of elephants), samples of ivory and control blocks of wood to different groups of related elephants. The results were interesting. When presented with skulls of elephants, rhinos and buffaloes, the elephants spent more time investigating the elephant skulls. So far so good, you might say. But when presented with three elephant skulls, one of which was from a former familial matriarch, the elephants showed no greater interest in familial bones than the bones of strangers. I wondered why this evidence wasn't more commonly known or cited. I had never heard about it. I wonder whether this information doesn't make it into most TV nature documentaries because it doesn't fit the narrative that elephants are incredible mourners and that they are just like us, which they may or may not turn out to be. They may mourn in ways we cannot imagine or understand.

Near the end of my tether, I decided to think of other ways to get my sample size up. Where could I find thousands of individuals of a cognitively advanced species, living together, being housed similarly and raised in the same repeatable conditions over and over again? I thought long and hard about this. In fact, I thought about it for many months. And then it hit me. Donkeys. Donkeys are kept in paddock after paddock in Britain, often rehomed in special sanctuaries. Donkeys often form pair bonds, unlike most livestock. They form very strong bonds with one another, I'd always been told. Surely sometimes one must die, leaving a partner on its own. Surely, donkeys would be where I would find some answers. Repeatable experiences. Trends. Something approaching evidence. I emailed The Donkey Sanctuary in Sidmouth, Devon, to see whether they would talk to me based on their experiences of looking after 3,000 donkeys spread across eight farms and caring for another 1,500 donkeys in private foster homes. What was their experience of animal mourning? I asked. They responded. Suzi Cretney, their PR manager, invited me over to meet with their head of research, Dr Faith Burden, to show me around. A week later I was on my way to Devon.

I know very little about donkeys. My beliefs about donkeys are pretty much the ones that have been fed to me through mainstream media: that donkeys are hard-working; that donkeys are stubborn; that donkeys are a bit world-weary – that sort of thing. As we sat in the Donkey Sanctuary cafe surrounded by paddock after paddock after paddock of donkeys, Suzi and Faith gave me a crash course in what donkeys were like in reality. And I realised I had got them all wrong. 'When you think about donkeys ...' said Faith, opening her can of soft drink, 'think about where they come from: try and remember that, inside each one, is a wild donkey. That's really important.' I jotted this down. 'I mean, they've only been domesticated for 6,000 years so, you know, you can't really take the desert out of them. It's a part of them,' she said. 'Most of what we see, hear and study about donkeys physiologically and behaviourally all comes back to that earlier life: that wild donkey in the deserts of Ethiopia.' I noted this too. 'They're not glamorous animals,' Suzi told me from across the table, 'but they're animals on the backs of which civilisations have been built. Literally: they built whole civilisations.'

I poured my tea. I quite liked that we had started off with this. In popular culture, there is a feeling that people that care for donkeys are somehow twee and eccentric. I was a tiny bit worried that Suzi and Faith would be actually very twee and eccentric. But they didn't seem to be. They seemed lovely. And scientific. Very rational people. And they were quick to understand the predicament I was in.

I explained about the problems I had in getting to grips with such an amorphous phenomenon as animal grief and mourning. I explained that I had almost ditched it from my research. 'Could donkeys offer me any insights in this?' I wondered. Suzi and Faith looked at one another, deciding politely who should answer. Suzi smiled at Faith, beckoning her to take it. I wondered if they had talked much about this subject before. 'Well,' Faith began. 'Donkeys are a contrary animal to understand, that's for sure. They're not like horses. There's no stallions or harems of mares. They're often either solitary or living in pairs. In some ways that probably makes them strong-bonding,' she said, looking at Suzi for some sort of reassurance. Suzi looked at me. 'They bond closely,' she said, paraphrasing a bit. 'They have to stick together. If there's only two of you, and one of you dies, you have as much invested in companionship as life alone. You'll notice it. It's a big deal.' 'How that translates in domestic situations is what we see here,' said Faith. 'Some donkeys here are so tightly bonded you can't separate them by a fence, let alone a stable.' 'So what happens when one of them dies?' I asked. There was a brief silence. Faith cut through it: 'Some donkeys respond to death with overt symptoms of distress, but not many, and by no means all. That bit's really important. Not every donkey behaves the same around dead donkeys.'

They gave me some examples. They talked of some donkeys going off food for long periods after the loss of a companion, braying and even trying to jump from their enclosures on some occasions. They talked of anxiety. Nervousness. But they also told me of the variations of behaviours and character types that donkeys have. They talked about the donkeys that don't express much at all after the death of a companion. They talked of some donkeys apparently ending up much happier alone.

What I found very interesting was that, with these behaviours post-death being so difficult to predict, the Donkey Sanctuary plays it safe. 'For instance, if a donkey is to be euthanised,' Faith told me, 'then the companion donkey is given an opportunity to be there too. This is standard practice at the Donkey Sanctuary.' I was impressed. It seemed nice that, even though they couldn't guarantee a donkey's response, they were prepared for the worst. 'That's what we do here and we always advise others to do the same with donkeys that have bonded companions,'

Faith said. 'I'm guessing you've seen many euthanasias in your time …' I commented. 'Yeah, quite a few,' she said. 'How does it work, with the remaining donkey watching euthanasia take place?' I asked. Faith looked quite sombre all of a sudden. 'We let them stand there for as long as needed, depending on the behaviours they exhibit.' She twiddled the ring-pull on her soft drink. 'Some of the behaviours that you might see are very striking.' 'Oh?' I said. 'The companion donkey might sniff them. They might try and get them up. They might bray. They might run away.' She paused then. 'It's hard not to anthropomorphise,' she said, looking at me, slightly embarrassed at her response. 'I know,' I said. 'I know, it's hard …' she continued. 'I'm a scientist, you know? But then …' She smiled a bit. '… Then, of course, there are those donkeys that just aren't bothered. They take one glance and give a look to the dead companion to almost say "*I never liked you anyway*".' We laugh a little at the thought.

I asked Suzi and Faith whether, and how often, they used the words '*grief*' or '*mourning*' to explain the behaviours they occasionally saw – particularly Suzi, who worked in PR, whose job it was to help frame these animals as creatures worth donating money to save. 'Scientifically, I'd label it as distress,' said Faith quite assertively. 'Yep, that's how I'd describe it. It's distress.' I looked at Suzi. Suzi thought about the question for a moment. 'In humans, regardless of religion, regardless of where you look, you'll see a marking of the passing,' she said. 'But … we're talking here about donkeys responding to the death of one another. Not family members, or anything like that. News of a death – the death itself – doesn't ripple throughout the herd. They're not all affected. And they're not all affected, all the time, even if it is their best friend,' she told me. As Suzi talked I could see Faith thinking more about the question. 'So, what's the correct terminology for you?' I asked. 'Distress. I think distress is a good word,' said Faith.

We all paused. There was a silence. It was the only silence in our entire conversation. I felt that distress really was the right word. Perfect, really. Nearly all humans know sadness and loss at losing someone close and many, unfairly, have felt this pain sharply numerous times (many of us feel it when we lose pets, too). Though it's hard for us to know truly the emotional states that people other than ourselves feel, I think it's a reasonable guess to say that you hurt like I do at funerals. That you cry. That you feel an intense weight of sadness. That it really hurts. You may, like me, have literally fallen over at the news of someone close to you dying. Literally, fallen. Make no mistake – for the bulk of us, these are loud and powerful emotions. And they're relatively universal, too, across human cultures. This is mourning. This is grief. The more I had researched this topic in animals, the more I came to feel that the words we choose really matter. There is no animal that responds as predictably to death as humans. In fact, no other animal appears to come close. Freely labelling other animals as grieving or mourning then threatens to belittle what is an incredibly intense period of emotional turmoil in human life. At its worst, it makes a mockery of one of the most interesting things about our species. So 'distress'? Yes, I'm happy with it. I realise that I am happy, like Suzi and Faith, with the word 'distress'.

As we finish off our lunch, Faith tells me that her plan, working with UK universities, is to document in greater detail the events that take place after death, investigating and isolating behavioural patterns that could be the basis of further study. I was really pleased to hear this. That's what this area of science needs – creative ways to find out more about what may or may not being going on. We shook hands warmly, agreeing to keep in touch. Suzi said I was free, if I wished, to wander around and go and look at the various donkeys scattered in the nearby paddocks, which I gladly did. Sometimes they came over to greet me and allowed me to give them a little scratch or a pat on the head. Sometimes they didn't.

A few more weeks passed. I decided to keep this chapter in the book. I mentioned earlier how difficult a chapter this had been to write. I think now that this was partly because I had wrongly sought to bracket animals into two lots – those that understand death and those that don't – and look for the science to justify the categorisation of each. I failed to appreciate one enormous thing: that an animal's cognitive ability to understand death could actually be a continuum. It could be that we're all on a spectrum of understanding death. It's a bit like rats and laughter. They say that rats emit a high-pitched laugh when they are tickled, but would a rat understand a 30-minute satire sketch? No. There's a spectrum there. It may turn out to be a bit like that with death: that animals feel something of what we feel surrounding death, but nowhere near the complete human experience (assuming, that is, we can define such a thing – something Alex Thornton had pinned me on).

But there was one other peculiar thing that I had noticed about animal mourning and animal grief. It was this: it struck me that this had been the only question I was asked about again and again when telling people that I was journeying into death. It was always the same thing: 'Oh, will you talk about animal mourning?' they'd ask, before reeling out the anecdotes about their pets and friends' pets. This question followed me around for months and months. It really did. Why? Why were we all so interested in this particular aspect of animal behaviour? Why is whether animals grieve or mourn of such interest to animal-lovers?

At the end of this chapter I'm left still wondering. Maybe it's nice to imagine that our animals might miss us? Perhaps.

But I wonder now if it might be something deeper. I can't help feeling that, deep down, it's simply horrifying for us to consider being the only creatures in the universe that can imagine what's going to happen to us after we die. The only ones aware of our fate. The only ones to sometimes question our fate and seek a more emotionally friendly alternative. I think, deep down, it's lonely being the only ones at the top. The only ones in the know. And that, probably, is what makes us most human. Our own label is defined by our distress.

CHAPTER SIXTEEN

Who Wants to Live Forever?

Deep within the recesses of University College London there are worms. Nematode worms that regularly have a UV light shone upon them. Normal nematode worms show up as blue under these lights; they slither across microscope slides in a perfect S, like heartbeats on an ECG. But occasionally something else happens to the worms – some of them suddenly seem to ignite. There appears a bright blue spark within their intestines. It spreads out across the worm's body, and a blue explosion washes over the worm like a forest fire consuming all in its way, leaving only charcoal – death – behind. This phenomenon is called 'death fluorescence'. Originally thought to be a marker that results from the ageing process, death fluorescence is now thought to be a straight-up marker of dying cells, and it is especially visible in these simple translucent nematode worms. Death fluorescence is true death, in other words.

It is mysterious and slightly eerie to watch videos of these worms being consumed in such a way. But it is also quite pretty – pretty because nematodes are really quite wonderful creatures. There are 25,000 named nematode species, though it's likely that there are a million or more species out there for us to discover, should we wish (sadly not many people wish this). Perhaps half of all nematode species are parasitic (which doesn't help matters). Nematodes are of particular interest to scientists because they are some of the most simple animals on Earth that have a tube-shaped body with a mouth at one end and an anus at the other. They are built like us, in other words, except much, much more simply. And within their genes we find programming instructions for how to build and regulate and care for that simple body plan. Nematode genes work, on the whole, in the same way that they work in our own bodies, thanks to the common ancestor that we shared with them about 600 million years ago. Their genes are our genes, mostly.

The world's most famous nematode – the fruit fly of worms if you will – is *Caenorhabditis elegans*. It is everything you want in a lab animal. See-through, free-living, easy to keep (it lives in the wild in soil), and scientists know its genome and the normal development of each of its 959 (in hermaphrodite females) and 1,031 (male) cells. *Caenorhabditis elegans* is tough. *C. elegans* specimens from the 2003 *Columbia* shuttle disaster were found to have survived after falling to Earth in a four-kilogram locker. That's how tough they are. Touchingly (if you like that sort of thing, which I do) their descendants were put into space by the shuttle *Endeavour.*

In earlier chapters I have explored the competing theories that may explain animal longevity, and I have discussed the impact of free radicals on animal cells when it comes to ageing. My research brought me into the realms of a host of animals that live longer than simple body size would predict. I came across cataract-ridden rockfish, wrote about naked mole rats defying all the odds, and I held in my hands Ming in its Perspex box – an animal that lived for more than half a millennium. However impressive this may sound, there is a problem with studying these creatures. In fact there is enormous difficulty in studying gerontology. All of these creatures make pretty poor study animals, because you have to wait 30 or 40 or 400 years just to see how they fare after experimental manipulation. As a result many scientists are rather turned off studying an animal that ages at the same speed as themselves. Smaller creatures are therefore more in vogue for gerontologists; scientists can obtain results much more quickly from something that lives days or weeks rather than years.

And so it is for this reason that *C. elegans* has become one of the favoured study creatures for this sort of thing. At this very moment, researchers are studying and observing and experimentally manipulating *C. elegans* in their tens of thousands, all over the world. For within them, say many scientists, are the mechanical workings of life and death. Within them are genetic mechanisms that can be picked apart like cogs and springs in an attempt to understand to a greater degree the causes of ageing and ultimately death … and, crucially, whether it can be fixed. These scientists are interested in a cure: a cure for ageing. A cure for age-related diseases. They're in it for the mother lode, essentially. As close as we might ever get to immortality …

But there is another reason that *C. elegans* hogs much of the ageing limelight. *C. elegans* can do something incredible. As it grows, this species can modify its lifespan depending on the environment. Starve *C. elegans* and they pull back from reproduction and invest in life. Feed

C. elegans and they reproduce as normal and promptly die after a matter of weeks. Starved individuals live up to three times longer than well-fed worms, which is a striking result, no matter what your opinion otherwise of nematode worms. What's particularly incredible – and what *C. elegans* allows us to see clearly – is that key sets of genes appear to broker the whole deal. Life or death – genes 'choose' based on how much food there is. They choose life if there's not much to go around. They choose death if there's plenty to eat and all the sex is done and dusted. These special genes are important: they are the so-called 'gerontogenes' that control ageing. Master them and we master ageing, say the scientists who study them.

That small periods of starvation (or calorie restriction) can serve to prolong life may not be news to you, of course. Diet fads, daytime TV shows and many, many, many lifestyle books (which will all sell a great deal better than this) may have told you about it in recent years, and it's fair to say that the basic principles may turn out to be true across many animal species. For instance, studies where rats have been fed 40 per cent less over their life show that they live 50 per cent longer than their siblings and they suffer less from the diseases of old age, too. The same findings appear to hold strong in some fish species and also in dogs (and yeast, interestingly).

Though good data are lacking in humans (partly for the reasons outlined above: namely, who wants to wait 90 years for results?) it's likely that calorie restriction will have at least some effect on humans, too, though what effect we can't say. Calorie restriction doesn't always seem to pay off. Wild mice, for instance, fail to show any increased longevity when their calories are restricted. Indeed, longevity studies on calorie-restricted non-human primates have had conflicting and quite confusing results. Either way, the gerontogenes – the genetic brokers of life and death – are of key interest to scientists. Gerontogenes have within them the potential for extended life. They could be used to make us live longer … if we want to live longer, that is.

Why calorie restriction appears to have such an impact on the lifespan of some creatures has been considered for decades. In the 1970s, the gerontologist Tom Kirkwood attacked the problem with particular gusto. Far from being about death, Kirkwood saw flexible longevity as an adaptation to do with sex. Though the discovery of gerontogenes was a long way off, Kirkwood made his argument based on economic grounds alone. He predicted that in situations where energy was limited, expensive investment in reproduction may become futile and that instead energy should be transferred into longevity, maintaining cells and sustaining life in the hope of good times ahead when the opportunities for sex reappear. This economic approach is something we have come to know only too well in recent years. Consider bankers. It's the same with them. In hard times bankers don't invest – they shrivel up and maintain their own interests. They sit and wait. They age. So, it turns out, it's probably the same with bodies. It is in the good times that they can afford to be more frivolous; they spread their influence and grow; they breed. They live faster lives. The vast majority of animals on Earth may be capable of the same trick as *C. elegans*.

But *C. elegans* is just one of a number of animals to be pulled apart by gerontologists in such a way; there are other animals that are similarly being picked apart by geneticists and gerontologists eager to expose other tricks organisms may be capable of to slow down death and prolong life. Two close competitors to *C. elegans* in the genetic search for the elixir of life are jellyfish and *Hydra*. Both, like *C. elegans*, are simple multicellular animals whose cell development and genetic information can be readily mapped. For both, opposing camps of scientists have gathered around them.

The so-called 'immortal jellyfish', *Turritopsis*, is perhaps the most lauded by those in the jellyfish camp. It is capable, they say, of ageing in reverse. Most jellyfish species begin life as tiny tentacle-covered polyps, glued to rocks, which then 'birth' free-swimming life stages (medusae) which will grow into adults, at which point they produce eggs and sperm and promptly die. But the immortal jellyfish flouts these normal jellyfish rules in the most remarkable way. The immortal jellyfish, as a post-spawning medusa, doesn't die. Under certain conditions it can sink to the seafloor, fold in on itself and take on the polyp form once again. It lives and lives again, in other words, while everything else lives and dies. Some people call it the Benjamin Button jellyfish for this very reason.

How the immortal jellyfish manages this is still up for grabs. They are difficult animals to study – immortal jellyfish are incredibly difficult to keep and breed in labs, for a start. The only scientist to do so for any significant period of time is Shin Kubota, a Kyoto University scientist who for the last 15 years has spent three hours each day caring for a brood of just 100 jellyfish. Kubota is a man wedded to an idea: '*Turritopsis* application for human beings is the most wonderful dream of mankind,' he told *The New York Times* in 2012. 'Once we determine how the jellyfish rejuvenates itself, we should achieve very great things.'

So there's *C. elegans* and there's the immortal jellyfish – both being poked and prodded and probed by research scientists eager to uncover their secrets to slow ageing. And then there are others. *Hydra*, a genus that belongs to the same group as jellyfish, is another important player in the anti-ageing scene. *Hydra* win points for being much more common than immortal jellyfish (they exist in most unpolluted freshwater lakes, ponds and rivers on Earth, predating passing food rather like anemones do), and because they can be cultured and manipulated in labs very easily. Since 1998, when the gerontologist Daniel Martinez declared to the world (in a paper in *Experimental Gerontology*) that '*Hydra* may have indeed escaped senescence and may be potentially immortal', many gerontologists have gathered around the *Hydra* camp, sticking with them even though other scientists have questioned their supposed miraculous ability to limit ageing. *Hydra* are of particular interest for the regenerative abilities of their stem cells (the body's master cells, capable of dividing into any one of hundreds of cell types). It seems that *Hydra* stem cells may have an unlimited capacity for self-renewal, which makes them different to most other creatures. This buys *Hydra* a place at the gerontology table, and then some.

Undoubtedly the interest in this trio of animals – nematodes, jellyfish and *Hydra* – is growing. And so, with them, grows the rhetoric about the impact that such studies could one day have on our own bodies. Anti-ageing is a real thing. It's a real movement. And it's only a matter of time before we can wield and manipulate lifespans more in line with what we wish, potentially living well over a century rather than mere decades. The question is when …

And then there is yeast. Yeast has become another classic study creature. Within yeast there has been discovered an unusual class of enzymes called sirtuins, which have been implicated in many of the cellular processes described in this book: apoptosis, stress resistance and the role of mitochondria in ageing. Incredibly, in 2014, by experimentally manipulating the sirtuin-like enzymes in mice, scientists managed to extend lifespan significantly and improve health in the process. The next step, one expects, is clinical trials; one day with people. This really is happening right now. The anti-ageing industry is evolving; an industry that is eyeing up the potential in enzymes and in our genes; a world where jellyfish and tiny worms unlock our wildest fantasies of living forever. Whether this amazes or terrifies you depends on your attitude to death, I guess. Me? I'm undecided. I wanted an opportunity to see this industry up close. I wanted to observe what it might look like, and where it might go, and the sorts of people that are enamoured with the idea of living longer. And that's when I saw it. An advert for an event like no other. The Anti-Ageing Show, taking place at London Olympia, was coming up the following week. So I bought tickets; I bought tickets and had my head blown off.

But before I tell you about what happened, I just want to reiterate something: I really meant it when I said that the anti-ageing industry is growing. According to *The Times*, in 2014 the global anti-ageing market reached $150 billion. Now it is approaching $200 billion. Where once facelifts were vogue, anti-ageing therapies now involve diets, exercise, hormones and all sorts of energising rubs and creams as well as the enzymes, the antioxidants and the possibilities that genetics throws into the mix, care of worms and jellyfish and, possibly, yeast. It really is big business. In September 2013, Google's CEO Larry Page announced Calico (the California Life Company). Calico's aim is to eliminate age-related diseases, like Alzheimer's and cancer, through a $1.5 billion life-extension research centre. Though some argue that Google are flag-flying for their 'Don't be evil' mantra (remember that?), there's a good chance they've spotted the market trend: in 2025, there will be twice the number of over-sixties as there were in 1995. The market is building itself, year on year, awaiting the products they are trying to produce.

Other institutions are also seeing the potential. There's the Age Reversal Fund ('Our plan is to achieve maximum advances in wellness, youthfulness, longevity and profits … in minimum time'). Then there's Larry Ellison, CEO of the tech company Oracle; the money man (according to *Forbes*) behind the Ellison Medical Foundation, another anti-ageing research centre. And there's also the SENS (Strategies for Engineered Negligible Senescence) Research Foundation, supported, in part, by Peter Thiel, co-founder of PayPal. There's no doubt about it – ageing is big business and for good reason: there's lives to be saved from the ravages of a host of age-related diseases, and there's bundles of money to be made. A planet of potential customers, few of whom seem ok with the idea of getting old and dying naturally. Anyway, on to the show …

Olympia, for those who have never had the pleasure on a warm day, is HOT. It's essentially a hangar-sized greenhouse-cum-oven. By the time I arrive, after a long walk from the Tube, I am a very sweaty man, completely overdressed, walking into a sauna. I arrive at about 10.30 am to find the show in full swing. The entrance going into the main hall is like an arrivals corridor in a country very much hotter than Britain. A gaggle of salespeople awaits us, handing us literature and free samples as we walk in; I am greeted with leaflets about facial acupuncture, 'bespoke advanced facials' and free samples of probiotic balancing cream which look and smell suspiciously like yoghurt. I flick open the exhibition magazine I have been handed and am met immediately with a half-page advert for something called 'gong therapy' where a man rings a gong and, to paraphrase, everyone feels much younger and better about things.

I decide it's best to sit down and have a coffee. I have decaffeinated coffee at events like this because I don't really like big crowds. Big crowds make me anxious; they make me nervy and a bit edgy. I need to sit down. I can't find a seat near the coffee stand so I sit on a set of big wide benches in front of an empty stage instead, and I look at the show guide. The Anti-Ageing Show magazine tells me about the stands that I should definitely spend time at. I scan through the company names: Aluminé, DermaNutri, Donna Bella New Look New You, EDM Therapy, Forever Living Products, iGrow Hair Laser Rejuvenation System, Maddisons Unique Gel, Queen of Oil, Rejuve Me, Megawhite, The Jane Plan. I am surprised to see that a popular cat rehoming charity is here, presumably because if we all live longer we might want access to cats for longer, too.

I scan the room. It is a place for beautiful people, none of whom I can estimate the age of. Many of the women wear tight black pencil skirts that go up past their navels. They are very busy, these women. They move fast. When they walk their hands move up and down in front of them like they're on a StairMaster. I can only see a handful of men. Each is toned. Tanned. Smiling with their teeth. Muscular. I feel a wave of insecurity as salespeople come toward me with clipboards and more free samples, but many of them steer away at the last minute as if they suddenly realise, by looking at my non-fashionable clothes, that I would probably prefer an early death. Maybe it's because I'm copiously sweating. Maybe it's that they spot that I have misapplied my face yoghurt.

The coffee here tastes awful. I feel really horrendous all of a sudden, sitting in front of the empty stage. A host of pencil-skirted women begin sitting down around me. They fill up the benches in preparation for a presentation that is about to take place. On the makeshift platform in front of us, two people arrange a dark leather dental chair. There are some cameras being set up and microphones are brought out. More people gather. A slim lady in a tight-fitting black dress is brought onto the stage and seats herself in the chair. Another woman, almost identically dressed, comes out from behind the stage and stands next to her. She has a microphone. A third lady comes out, this time in a tight-fitting red dress. All of these women have exactly the same kind of bodies. It's like they've come out of a box. A man with gorgeous hair and a dazzling smile comes out next. He kind of … zings … onto the stage. He tests his mic, then smiles to the audience and introduces his team before beginning a presentation about his company's range of 'dermal fillers'.

I sit through their presentation patiently. It's actually a very professional pitch. As it goes on I notice that more and more people turn up to stop and listen. By the end there is a bustling crowd. The presentation ends, and the gentleman invites the whole audience onto the stage to watch up close the lady in the chair have her face injected with fillers and other assorted 'rejuvenation sugars'. I assume no one would want to see this but I am wrong; the stage wobbles as almost every single person in the audience floods onto the stage for a closer look at the minor surgery that the poor woman on the chair is about to go through.

I decide that this is not for me. My disgusting coffee finished, I take this as my cue to have a walk around. In the dead zone between youth and agedness nobody here knows quite how to sell to me. Some try to attract me in by asking questions about my wife – when her birthday is, what I plan to get her for her birthday, and how I should celebrate it with a surprise photoshoot for her or a spa treatment or some gong therapy. I mumble and stutter my way out of these encounters, looking increasingly sweaty and like the sort of person you'd want to avoid on the next lap of the show, which they then do.

I really am not good in crowds. In my heavy coffee-drinking years I have had a number of panic attacks, a couple of which were rather intense. Many of these attacks involved crowds and I am sure they were driven partly by the fear that some situations would be mortally embarrassing to have a panic attack in, a thought which would promptly then give me a panic attack. I know others who have the same problem. The Tube, job interviews, sitting in the middle of a row in the theatre: these are the sorts of places it would happen. The thought strikes me that this would be a horrendous venue for a panic attack. I immediately try to crush this thought. It lives now. The sweat washes the yoghurt off my face and onto my collar. I start wondering whether the barista accidently gave me caffeinated coffee just now. It must have been caffeinated, I think. It must have been. I hadn't looked at whether he'd ticked the little box on the cup that says 'DECAF' (maybe he didn't tick the little box on the cup that says 'DECAF'?). I spin a tiny bit. I look for another place to sit down and spot another empty stage. A big banner hangs over it: 'THE GRACE KELLY STAGE' it says. I sit down and enjoy being left alone for a few minutes.

I had come to the show wanting and rather expecting to immerse myself in the consumer side of the global anti-ageing industry and to see, maybe, whether the biology that I have been researching is nudging its way into the anti-ageing limelight. It seems that, on the whole, it isn't. Looking around there is no mention of mitochondria, for instance. No talk of free radicals. Nothing about *Hydra* or jellyfish or genetics or … well, gerontology at all. This struck me as odd, because the word *NATURAL* seems to feature on everything. Do these organisations not yet know of the potential for anti-ageing that lies within our DNA or within the worms or the birds or the bats or the jellyfish? It seems that, on the whole, they don't. A few people gather on the benches around me. Once again I flick through the free Anti-Ageing Show magazine. I keep seeing pictures of a Harley Street doctor called Dr Aamer Khan. He's on everything here. The leaflets, the banners, the website. In every shot he stands with arms crossed, with one hand suggestively on his chin. He has incredibly white teeth and looks intently professional and handsome. He has charisma. In fact many of the speakers at the show seem to have oodles of charisma just like him. People seem drawn to hearing them speak here, drawn to wanting to know what they know. Eager to hear from them the secret of looking beautiful for as long as possible.

Weirdly, I have noticed these charismatic personas in some of my research, too. Gerontologists are rather sexy and charismatic people, it seems. Take the immortal jellyfish scientist Shin Kubota, for instance. Kubota isn't only famous for his jellyfish lab. Kubota has an alter ego. In Japan, Kubota is also known as Mr Immortal Jellyfish Man – he has a costume and everything. Mr Immortal Jellyfish Man has released a number of pop songs and six albums in Japan. He is charismatic. And there are others like Kubota. Throughout my research I have come across pictures and interviews with a gerontologist called Aubrey de Grey (de Grey runs SENS), who likes to say that the first person to reach 1,000 has already been born. This is mentioned in many magazine and newspaper interviews. De Grey is charismatic and distinctive; he has a long pointed ginger beard, and in all his photos he looks intensely serious about things. De Grey photographs very well. These are famous gerontologists; experts and professionals who really know their stuff. People put a lot of trust in people like these.

The Grace Kelly stage starts filling up around me for a talk. As before, microphones and cameras are set up on the stage. The same familiar dentist's chair appears. I decide to stay for this talk too. It is about something called Dracula Therapy. A tall, serious-looking man with a chequered shirt jumps onto the stage and stands in front of us proudly. In his hands is a vial of blood plasma that he has recently taken from the woman who has come to recline on the shiny black leather dentist's chair in front of us. Barely giving us any warning at all, he proceeds to inject the blood plasma back into her body, deep into her muscle tissue. 'We have to inject near the bone,' he says to us over the loudspeaker, with a hard-to-pinpoint European accent. 'Not toooooo near the bone,' he says calmly. 'But deep ...

deeeeeeeeep … near the bone …' He concentrates, the needle going in deeper. It goes in as far as it will go, in fact. I wince and turn my head. The people to my right appear fine with what's going on; they lean in for a closer look. They love it. The lady next to me is actually eating an egg mayonnaise sandwich while watching. I wait a few moments and try to look again. He is now injecting the needle deep into her neck. I really don't like this. Incredibly, the lady is barely flinching. I don't like it at all. He pushes in deeper with the needle. Horrified, I flick my head to one side once again, involuntarily jerking my body so violently that … that …

And that was when it happened. The day took a turn for the worse. Things got bad. A fire like lightning ripped across my chest. A bolt of electricity. The whole front half of my body went immediately into an agonising spasm, from my neck and down into my torso, causing me, I was sure, imminently to die. I let out a low 'UHHHH' from deep within my body and momentarily pull a face like I am having a very painful and prolonged orgasm. The lady next to me stops eating her sandwich. 'You ok?' she asks dryly. 'I …' I let out another strange noise through gritted teeth. It sounds like I am about to cough something up. 'I … need to … UHHHHHHHH.' I stand up and swivel round, wincing, trying to get to my feet. 'I need to make an … UHHHHHHH,' I say. She carries on with her sandwich and goes back to watching the Dracula Therapy.

The pain is intense but I recognise it. I know this pain because I've had this before; it's a kind of neck spasm that refers pain right down the front left-hand side of my chest. It has every single hallmark of a heart attack but it isn't one. I haven't had one in four years. This one is sharper, though, like a hot pipe being stabbed into my chest and slowly rotated. I walk slowly, almost limping with the pain, my head locked, my neck at a weird angle. Every breath burns so I have to take quick shallow breaths as I walk. I must get out of here, I think. As I walk across the show I learn that it hurts slightly less when I say 'UHHHHH' when I exhale. I start saying 'UHHHHH' as I exhale. It helps. And so, in the middle of an Anti-Ageing Show, I become a man hobbling around with hunched shoulders and a grimacing face going 'uhhh … uhhh … uhhh …', on the verge of death.

I carry on regardless. I'll be alright, I think. I spot the signs of a panic attack coming on. I reassure myself that it probably isn't a heart attack. Probably. I think about asking for help (I can't tell you how annoying it is to be surrounded by people in white coats and not one of them is a proper bloody doctor). I decide to leave the show. I keep my head down and limp through the aisles of strange anti-ageing exhibitors. I walk past the people putting their heads into strange metal helmets attempting to regrow lost hair. I walk past a cubicle of women happily putting their mouths over a strange luminescent ball on the end of a plastic stick. I walk past the people lying in black leather dentist's chairs having their cheeks vibrated by strange probes and ladies having their lower halves covered in gel and yoghurt. I walk past the vibrating platforms. The tarot readers. The Cats Protection League. And then, finally, I make it out. Out into the fresh air; back into the world I had known before. A world where the normal rules of nature seem to play out. A world where death is natural and, on the whole, all things accept it. It took a week to recover.

Being at the show affected me more than I had expected. I was amazed that jellyfish, free radicals and calorie restriction hadn't featured much. This worried me a little, like I had witnessed the calm before the storm. I couldn't help but feel that it might only be a matter of time before the cosmetics industry is swamped by the potential of cellular and genetic therapies that we are now discovering and that I discovered earlier in my research into death.

For some reason I felt a bit of deep dread at the thought of this, which is strange because there are so many opportunities for these therapies to improve the world. It's incredible that we happen to be alive at a time when the diseases of old age could be realistically tackled and that we could live for longer than our bodies ever evolved to. Incredible that we happen to be alive at a time where global healthcare could be revolutionised; where diseases of old age become a thing of the past. So why was I feeling dread? Why had I essentially freaked out? I don't know. I can't really explain it. I guess all new and exploding areas of science fill the popular imagination with fear like this. Nuclear technology bred Godzilla; the IT revolution bred Terminator; the prospect of genetic engineering bred Jurassic Park. Weirdly, though, I don't see any public concern about the anti-ageing revolution which we are undoubtedly heading toward. I find this strange. It seems to be creeping up on us. I can't help but think that we're sleepwalking into a world where people can pay to live longer than others, and that feels strange. Unethical somehow. There are questions I have that no one can really answer because there are no answers yet. For instance, will such anti-ageing treatments be available free through healthcare plans or the National Health Service? If not, how will we make it so that the poor don't end up paying for the pensions of the age-endowed? How will we go about avoiding the creation of a two-tier society of late-lifers and early-lifers? I'm not sure anyone knows yet, and that's unsettling to me.

On the flip side, though, there seem to be such positives about the anti-ageing movement. It could be a great thing: imagine the billions of pounds of healthcare provision that could be saved if we tackled age-related diseases by tackling ageing itself. It could be incredible. In Britain, many argue that the NHS is in real danger of disappearing forever. Perhaps this could save it? So yes, it is exciting. And amazing. But then … still the doubt lingers in my mind. Does the world really need humans that live, and consume, for longer than they do currently? What happens if humans live regularly until they are 120? Will that be enough, or will the Silicon Valley elite still strive to push us to 125? Or 130? The truth is that we will always want to live longer. I'm sure of it. Always. For many people life is simply too wonderful to be happy with it ending. And that's a problem. A big problem. Because it will.

CHAPTER SEVENTEEN

No, This is a Dead Frog

Later, long after the Anti-Ageing Show, I discover that there is a stumbling block to Aubrey de Grey's notion that some humans born today may be the first to live for a thousand years or more. That problem is the brain. Or more specifically, that problem is the neurons within our brain. For neurons don't replicate or regenerate in the same manner as normal cells. And if they could be made to replicate, each would lose their 10,000 synaptic connections to other neurons; synapses that together make up the thing you and I call experiences and memories. Lose them and we lose ourselves ('And so the price of immortality is our humanity,' concludes Nick Lane in *Life Ascending*).

Maybe de Grey is right, though: if we can tackle bodily ageing, maybe in time we will learn to tackle ageing in the brain, too, just in ways we cannot yet imagine. It seems unfathomable, but then so did so much of gerontology 30 years ago. When considering ageing and the brain, we find ourselves on another scientific frontier: that of neurogenesis, the process by which new neurons are born. Here, the worms and the jellyfish can't really help – it is birds, and their songs, that offer us a glimpse of the impossible. But first, we must revisit what we were all told about neurons …

You know the story. We were all taught it in school – we were told that the brain is fully kitted out with neurons shortly after birth, which are then trimmed back over the early years of life and on into later life. We were taught this because it really was the popular thinking about brains for decades, until someone thought to look into the brains of birds and they realised that it was actually quite, but not totally, wrong. The birds proved it. Far from being a static module incapable of new growth, brains of some male songbirds (at least the parts associated with song) appear to shrink back in summer (as territorial conflict and sexual tensions ease), only to 'regrow' in autumn in time to learn and rehearse their songs for the following spring. Essentially, parts of their brains could die and were able to grow back. After thousands of cells died, thousands more were born. This was a genuinely ground-breaking discovery. The research, undertaken by Steven Goldman and Fernando Nottebohm of the Rockefeller University, buoyed the concept of neurogenesis and inspired other zoologists and neurologists to seek out more such cases in nature. And they found them. Marmosets, rats, tree shrews, rabbits – all seemed capable of small amounts of neurogenesis.

In 1998, humans joined this elite pack. It turned out that our brains are also capable of growth, at least in the hippocampus – the part of the brain that, interestingly, is involved in memory and information storage. In humans, what might these new neurons do? That's the big question. Do they offer extra space for memory as bodies grow old, or is this just an elaborate developmental glitch, a cosmic red herring somehow? Predictably, scientists appear split over this point. Natural selection hates waste, and wasted cells are expensive, so these regrown cells must have a purpose, surely, but … there's also a chance that this is a side-avenue which has little consequence or bearing on the way in which brains work at all.

One thing remains clear, though: birds that cut back and later regrow the parts of their brain associated with song have to relearn that song. Their songs don't come back fully formed. Their experience is extinguished. No matter how hard we try, the human 'YOU' may never live longer than your neurons will allow. So how long is that, you may ask. The truth is that once more (and I know this is getting tiresome) we don't know. In 2013, Italian neurosurgeons discovered that mouse neurons transplanted into a rat's brain didn't die after 18 months, which is the normal lifespan of a mouse. In fact the mouse neurons continued working fine in the rat – these neurons were still alive and kicking in rats living twice as long. Mouse brains live for longer than anyone ever gave them credit.

There's a chance it may be the same for our own brains. Dr Lorenzo Magrassi (the co-author of the research on mouse neurons outlined above) thinks 160 might not be a problem. Others disagree. Nick Lane, for instance, suspects 120 years may be the upper limit for the lifespan of our neurons. Whether our children will find this out for themselves is still anyone's guess. Either way, even the most enthusiastic (and most economically endowed) of life-extenders will still have to face up to that uncomfortable truth. You are your brain. As with birds, if you refresh your brain cells your tune will change. And you may lose yourself in the process.

Driving up the motorway on my way to Alison's house I spend rather a great deal of time deliberating whether we should shake hands when we meet again or whether we should hug. Although I've only met Alison twice (probably a handshake then) she has been really supportive in her role as deathsplainer during my journey. Through regular correspondence she has provided me with such things as contacts (she got me in with Peter to talk dead pigs), advice ('this is a dead frog') and dead magpies (dead magpies). Plus, with her showing me the ermine moths and their incredible silken construction, I got to see first-hand the splendour with which nature acts, particularly when parasites force life to dangle closely over the edge. And through the ermines I also got to see the misinformed barbaric acts that humanity is capable of, when thoughts of death are invoked. I owe Alison a great deal, actually.

When I arrive and when we finally meet we mutually opt for a hug, rather than a handshake (though I suspect this is partly the Canadian in her). Alison has a lovely house. I like looking in people's houses and at Alison's I continue this passion with aplomb. She and her partner have made this personal space a paradise. Everything is in bloom, inside and out. There are plants everywhere. As we sit and chat in the kitchen, blue tits and sparrows gossip on the garden fence, making fleeting forays back and forth to well-stocked feeders. There are seedlings growing on a drying rack on the draining board. Binoculars on the kitchen table. A large poster of a dragonfly, lifted straight out of a Victorian identification guide, covers a wall at the bottom of the stairs. At the bottom of the garden is a large homemade solar array which her partner has designed and built. There are bee nest-boxes. Bird boxes. More feeders. Trees. Green. The works. Alison is much more into life than I had given her credit for.

'So ...' I ask with a grin. 'Any dead birds in the house?' She looks at me as if I've just asked the stupidest question in the world. 'Oh yes, course,' she says matter-of-factly. 'Of course. There's a dead blackbird drying out round the back.' We gather up our things and walk out to get a coffee somewhere. I decided a few weeks before, not long after the Anti-Ageing Show, to make the journey back up to see Alison in Birchwood after she started making subtle references on Twitter to the ermine caterpillars and the lost trees. 'OH MY GOD, ARE THEY BACK?' I'd asked. Then I had thought about it. I mailed Alison again straight away. 'ACTUALLY, DON'T TELL ME! I WANT IT TO BE A SURPRISE! I'M GOING TO COME AND VISIT!' I suspected that somehow, against all odds, the ermines were going to come back. I thought it would be a really fitting end to the book – a total cliché really – to have it finish off with me seeing the ermine caterpillars again and the trees on which they once feasted resurrected from the dead. A metaphor for how, no matter how hard we try not to think about it, life sprouts once more from out of the ashes of death. Perfect, I'd thought.

And so here I was, back in Birchwood. As we headed off to get our coffee, Alison took us on a detour so that we could go past the site of last year's chaos. We walked for 10 minutes or so through and along the twisting paths over roads and between streets. A minute later Alison stopped suddenly and looked at me expectantly. 'Look around,' she said. I stopped. We were standing in the middle of a wide-open pedestrian walkway with rows of terraced houses to our left and to our right. 'This?' I said, slowly coming to terms with where we were stood. 'This … was it?' I squeaked. 'This is where the trees were?' 'Yep,' said Alison gravely. 'This is the walkway where the trees were.' It looked totally different. It seemed so open and light and sunny and … not green in any way. Just brightly lit tarmac and bricks really.

Alison pointed to the stumps. There they were. Four neat little tree stumps in what was once an avenue. Each stump had about a 30cm girth. I approached the nearest one and tried to count the rings but they had sadly weathered away. The stumps felt terribly wet. Waterlogged somehow. They were starting to rot. 'Come over here,' called Alison from up the street. 'Look at this one …' She stood in front of a stump which had a tiny sprig coming out of it. From out of the sprig a tiny leaf had sprouted. This single leaf almost seemed to suckle on the sun's rays that beat down upon it. That single leaf, powering a whole tree stump now. A tree with delusions of grandeur. I inspected the other tree trunks. Nothing. No other leaves at all. Just stumps. I was a tiny bit downcast at the sight of what Alison had brought me to inspect: an empty street with a single green leaf. No ermine caterpillars anywhere. I had misread Alison's communications. I realised my brilliant metaphor was bust. I spent about five minutes trying to take an arty photo of that single twig with a leaf coming out of it, but that didn't work out, either.

After a while I noticed that Alison had walked off down the street toward the coffee shop. I ran to catch her up, but she suddenly stopped before I got to her. She stopped underneath another tree. As I jogged toward her she turned round to me and pointed a single finger upwards into the branches above her head. An enormous smile crossed her face. Bloody hell, I thought. I caught her up and there it was. A single bird cherry tree had been forgotten about by the men with chainsaws. Within the branches above our heads were six apple-sized pockets of silk from which tiny caterpillars were due to emerge. A forgotten nook now bred life again. The caterpillars, like my metaphor, had pulled themselves from out of the ashes. I cannot tell you how amazing it was. It was a splendid thing to have seen. Splendid. We agreed to keep news about the ermines quiet, in case the housing association should fire up the chainsaws once more. I couldn't stop smiling about it, though. I smiled and smiled and smiled.

At the coffee shop I unloaded a little bit on Alison about everything I'd seen and researched and everyone I'd talked to on my journey into death. The spiders, the spider scare stories, the pigs, the ants, distressed donkeys, African penguins, snowy owls, the grotto salamander, the red kites. The lot. I talked about how, over months and months and months, I had discovered that humans were closer than ever to having the keys to the city: close to having something nearer to eternal life, closer to immortality. At the very least, I explained breathlessly, we could realistically expect to see lifespan altered artificially in our own lifetimes. Some of us may live healthy lives right into old age, I explained. 'The first person to live to 1,000 may already have been born,' I relayed quite proudly.

But I also expressed a tinge of sadness. I talked with her about my panic attack at the Anti-Ageing Show and how strange it had all been. 'By the end of my journey, I… I guess I got a little bit fed up with going on and on about death …' I admitted, honestly. Death really had taken me over for a while. Mourning, grieving, consciousness; our desperate attempts at holding out for just a few years longer. The scare stories. The extinction stories. The Dracula Therapy. It all got … it all got a bit much, I explained. Alison nodded. 'I get that feeling too sometimes,' she said. She paused, sipping her coffee. 'I don't know if everyone who studies death gets it, but I think you reach a point where you have to find a balance. I study death – the science of death, the societal and cultural interactions of death – but my everyday life is surrounded by, well, life,' she said a bit more perkily. I'd noticed that about her. She continued. 'Yes, I research death,' she agreed. 'But I'm much more interested in what that tells us about life.'

I mentioned to Alison about the leaf-cutting ants and how they fling the corpses of their nest-mates off the bottom of the nest into the chasm below. We talked about funeral plans. I asked her if she had any requests for what happens to her body after death. 'Ohhhh, I'm very much a supporter of green burials,' she said. 'You know, put me in a hessian bag, stick me in a nice field with some trees and some flowers. That sort of thing.' I asked her if she was ok with the mental image of her body being pulled apart by invertebrates. 'It's fine if animals want to eat me.' She shrugged. 'I don't have a sense of attachment to my remains. I would feel much more comfortable knowing that, in 50 years' time, my body has become a field of flowers,' she said quite breezily. I thought about this. 'I quite like the idea of a tree growing out of my head,' I said quietly. 'I quite like the idea of a horse chestnut tree's roots pulling apart my skeleton, draining my atoms from the soil to make conkers that my great grandkids can collect and keep on their windowsills.'

But, even still, as I said this to Alison, something deep within me coiled up when I pictured my decaying body lying there in the soil. Embarrassed, I told Alison about my horror at imagining my own body lying there. 'The idea of it …' I heard myself utter. 'The idea of my body lying there … *decomposing*. I don't know, it just makes me … uncomfortable,' I said. 'I just … I just don't like the thought of it. It's irrational, I know.' Alison smiled. 'Well, we're all irrational when it comes to death,' she said. We looked out of the window. 'I'm irrational too. The thing that I consider all the time is that I'm a Canadian living in a country where I'm not from. Ideally, if I die, I'd like to be buried where I'm from. But I probably won't be.' This sounded strange coming from a death professional like Alison. 'Why?' I asked. 'Why would that matter to you? After all, you'll be dead, right?' This came out much more bluntly than I had anticipated, so I immediately rephrased it. 'It's strange that you care,' I said. 'I do care …' said Alison, pausing for a moment. 'I care because I'm leaving the people I care about behind.' We finished up our coffees.

'We're seriously messed up, aren't we?' I said to Alison on the walk back. 'As a species, I mean. We have some serious issues about death, don't you think?' 'Go on …' she said. 'Well,' I continued. 'I've been messing about with this bloody book for so long and I can't help but feel that we have such trouble grasping the whole concept of it. We understand death on paper, sure, but … but we struggle.' We talked about my experience with the whole 'mourning' label and how people had often been very interested in my opinion of whether or not animals grieve or mourn. I explained my unprovable theory that actually I think we're quite lonely sometimes being the only animal to rationally and realistically be able to ponder our own mortality, and that this loneliness comes out in all sorts of weird ways. I expressed to Alison my belief that the ermine moth caterpillars had been victims of our itchy trigger finger when it comes to death; that they reminded us of death so much that they had freaked us out and they had had to go. We talked about the spiders and maggots.

In fact, the more we walked, the more I think all of my thoughts just seemed to somehow pour out of me. It turned into a bit of a diatribe, and looking back I am very sorry that Alison had to hear it all without me having time to organise my thoughts a little more. I couldn't stop talking about death. 'Death strips our self-worth from us,' I said at one point. 'It makes our lives meaningless. I think the weirdness we have about death is an emergent property of our impressive cognition, and that sometimes we struggle with that and that's why death terrifies us.'

Alison listened patiently and then she offered up her own opinion on death, which was far more succinct than my own. 'If we didn't have this weirdness surrounding death,' she said, 'then we might be totally different as a society.' She paused for a few moments, making sure I took it in. 'I mean, if we didn't give a shit about death, then …' She raised her eyebrows, and shook her head. 'I don't know … that's what being human is all about, isn't it?' Ahh. Perhaps it really is as simple as that, I thought. Perhaps the best expression of being human and being alive in modern times is to be a bit weird about death. I think

Alison was totally right. In a funny sort of way, I have come to really like that about many humans that I know. Many of my friends and family are very irrational, and completely spooked by their own mortality, and I wouldn't want to change that. I wouldn't want to change that … much.

Epilogue: The Meaning of the Loa Loa

The extraordinary noise caused by the horses' hoofs makes the fishes issue from the mud, and excites them into combat. These yellowish and livid eels, resembling large aquatic snakes, swim at the surface of the water, and crowd under the bellies of the horses and mules. The struggle between animals of so different an organisation affords a very interesting sight …

So wrote Alexander von Humboldt in his account of his 1799 to 1804 travels in South America.

The eels, stunned by the noise, defend themselves by repeated discharges of their electrical batteries, and for a long time seem likely to obtain the victory. Several horses sink under the violence of the

> *invisible blows which they receive in the organs most essential to life, and, benumbed by the force and frequency of the shocks, disappear beneath the surface. Others, panting, with erect mane and haggard eyes expressive of anguish, raise themselves, and endeavour to escape from the storm which overtakes them.*

Von Humboldt's words are graphic and almost haunting in their brutality. There is a dramatic style about these sentences which Darwin himself (in *The Voyage of the Beagle*) is said to have emulated. His observations of electric eels doesn't end there, either. On the fate of the horses he continues:

> *A few, however,* [...] *gain the shore, stumble at every step and stretch themselves out on the sand, exhausted with fatigue, and having their limbs benumbed by the electric shocks of the gymnoti* [eels]. *In less than five minutes two horses were killed. The eel, which is five feet long, presses itself against the belly of the horse, and makes a discharge along the whole extent of its electric organ. It attacks at once the heart, the viscera, and the cœliac plexus of the abdominal nerves. It is natural that the effect which a horse experiences should be more powerful than that produced by the same fish on man, when he touches it only by one of the extremities. The horses are probably not killed, but only stunned; they are drowned from the impossibility of rising amid the prolonged struggle between the other horses and eels.*

Nature is savage. Savage and brutal and devastating in its apparent efficiency. Brutal in its chaos. Brutal, gory, visceral. Animals die by strangulation, through the ingestion of poisons, via venomous fangs and spines, through suffocation, starvation and evisceration; the fact that electrocution finds itself on the list of ways to die really comes as little surprise. But humans are brutal too. For the electrocution of horses described above was a staged act, undertaken by Indians to show von Humboldt, at his request, how the locals fished for eels. Once the electric eels had lost their charge after wasting it all on the horses, the fishermen entered the water and pulled them out with their hands. Simple as that. Nature is brutal, but more brutal are the humans that enact such horrors knowingly.

I am aware that suffering and death, whatever the cause, is a problem for many nature-lovers on a personal, religious and philosophical level. So gruesome can be the reality of nature that many, including Darwin, have seen their faith challenged through observations of the apparent pernicious barbarism that abounds there. In an 1860 letter to the American naturalist Asa Gray, Darwin famously questioned how an all-knowing and caring God would create parasitic ichneumonid wasps. He wrote: 'There seems to me too much misery in the world … I cannot persuade myself that a beneficent and omnipotent God would have designedly created the Ichneumonidae with the express intention of their feeding within the living bodies of caterpillars, or that a cat should play with mice.'

David Attenborough has his own version of the ichneumonid: the loa loa – a nematode transmitted to humans through fly bites. The worm makes its way through connective tissue and occasionally drills its way out of eyeballs. The loa loa has become Attenborough's philosophical weapon of choice when attacked by creationists for failing to credit God with his documentaries' many natural wonders. 'What God would create a loa loa?' Sir David asks with (what I imagine to be) a patient and good-natured grin.

I'm not sure where I stand on God now that my book is finished. I have never been very good at debating with others God's existence, to be honest with you. I get too sweaty when particularly enthusiastic religious people pepper me with probing philosophical questions and, often, an enormous big itchy rash creeps up my neck which everyone seems to spend a great deal of time staring at, which only serves to make it worse and even more itchy. I occasionally find myself bringing up the thorny subject of the loa loa in those discussions that I do have. From scientifically minded religious people I am often met with a response that natural selection is God's process for getting things moving on Earth. A kind of cosmic terraforming event He has planned to prepare a world in which we humans can live, even if it does take billions of years to really get going. To subscribers of this particular worldview, nematode worms, flies that infest livestock with their maggots, cloaca-burrowing mites and ichneumon wasps are simply emergent noise in this God machine. Software bugs (and sometimes literally they are bugs) that can't be removed without the whole system breaking down, which would expose the mind of the Creator who is desperate to remain hidden to make us show faith that he exists as a way to prove to Him that we all love him which is actually a very complicated and strangely self-conscious thing for Him to want from us that I've never understood.

Anyway. For me, in my early academic years, the existence of such creatures as the loa loa did make me question any faith I once had. But perhaps, for a while, I went too far the other way. I relished the brutal acts of nature as a means to display openly to others what I believed was the type of world in which we live. Tennyson's 'red in tooth and claw' became, for me, a Darwinian war cry. It was a badge of honour; a way to fit in with the scientific reality in which I had found myself in my twenties. A way to fit in with fellow academics, too.

But now, over the years and after writing this book, I have softened a little. For me now, nature isn't a brutal place. It isn't only a place red in tooth and claw. It's not only a place of death. To me nature has become a place of wild and unthinking possibility; a maker of diversity, variation and incessant creation. And death? Death is the process through which more life is created. New species owe their life to it. Existing species are sustained by it.

Extinct species are remembered by it. Schrödinger asked 'What is Life?' With death, the answer seems quite simple to me now. Mostly death is what you make of it; an understanding through which we might make better use of the time we have.

One of my many worries about this book being published now that I am at the end is that someone accuses me of downplaying the nastiness and the suffering and the pain and the gore from which we, and all things, have sprung; the realities of which play out each day. Gore. Blood. Entrails. Extinction. Cancer. Pain. Loss. I have mentioned all these things in this book, but perhaps not enough. Conversely, you might have noticed how many times parasites have been mentioned. Why? Because we are so often troubled by the thought of them, as Darwin was. What Darwin didn't appreciate is that we are all parasites. Parasites of life. Parasites within food webs. Parasites of the sun or of deep-sea ocean vents. We all take stuff, make stuff, and give it back in a more accessible chaotic doggy-bag for the universe in a wide range of forms: hot air, through our faeces, or through our own stinking corpses when we fail to balance the gerontological books properly. We are all agents of chaos, equal in our aims. All of us. And that's worth celebrating.

Occasionally I am invited to give talks to audiences about evolution and how natural selection works. I babble on about this and that whilst showing slides of the Great Tree of Life; I talk about slugs, jellyfish, fruit bats, duck genitalia, peacock spiders, deep-sea wars between giant squid and sperm whales, and I talk about how lucky we are to be alive at a time where all these wonders can be studied and understood. I talk about how much we still have to understand about nature; the colours and songs of non-bird dinosaurs, for instance, the last universal common ancestor of all living things, where and how slug mites have sex, and the many unusual things to do with ageing outlined in this book. And then I talk about how far we've come as a grassland-adapted primate that came (mostly) good; a grassland ape that became us. Human.

Once, at the end of a talk, a young boy came up to me, looking rather solemn. The Tree of Life was up on the presentation screen. He looked at it and quietly asked me something. He mumbled it. 'What's the point?' he said. I couldn't quite hear … 'Sorry?' I said. 'What's the meaning of it all?' he asked again. I stuttered and coughed and bumbled on about natural selection and this and that and, well, totally failed to give a good enough response and the young boy walked away with a bit of a weak smile. I have always regretted not giving a better answer. Now, finishing this book I realise what I should have said. He asked where the meaning was in life and I should have said that he was one of the most privileged human beings on the planet because, unlike most, he gets to find the meaning. He gets to choose the meaning. He gets to do something meaningful with his life. I think if we all found meaning in our lives, we'd probably find ourselves dying a bit better. For me, science gives me that meaning. And I am now planning for a good death.

Human life is too short. For many of us it will always be too short. For many of us it will always be too short no matter how much we manage to manipulate it and no matter how much money we throw at the problem. That problem will always remain: where is the meaning? I hope you can choose it well. I really hope you find out.

Acknowledgements

Almost every word of every sentence that appears in this book is, in some way, down to someone else; particularly the scientists and 'death professionals' offering their time, allowing me to be part of their interesting and important research, but also those who helped in other ways through direction and comments on the text. I am enormously grateful to them all for their time and they are named here together. They are (in no particular order): Rhiannon Bates and Dave Clarke (ZSL, London Zoo), Carla Valentine (Barts Pathology Museum), Megan Rosenbloom (Death Salon), Paul Butler, Peter Cross, Louisa Preston, Becky Wragg Sykes, Anne Hilborn, Jonathan Green, Sue Armstrong, Ben Hoare, Alan Stubbs, Jolle Jolles, Melissa Harrison, Bruna Bezerra, John Walters, Paul Stancliffe, Simon Leather, John Hutchinson, Andrew Whitehouse and Jo Gilvear (Buglife), Chris Cathrine, Chris Faulkes, Alex Thornton, Andrew Dawes, Lloyd and Rose Buck, Adam Hart and Stace Fairhurst, Faith Burden and Suzi Cretney, Ben Garrod, Ed Yong, Matthew Cobb, Darren Naish, Kane Brides, Phil 'Lash' Ashton, Ben and Jane Barlow, Lawrence Foster, Katherine Allen, Kathy Wormald, Silviu Petrovan, Matthew Oates, Richard Jones, Richard Fox, Amin Khan (www.aminart.co.uk) and Marsha Day (and everyone at Clipston Book Club!). And 'John' of false widow spider fame. And, of course, an enormous thank you to my deathsplainer, Alison Atkin.

This book couldn't have come about without Nick Lane's fantastic books, *Oxygen: The Molecule That Made the World* (2002), *Power, Sex, Suicide: Mitochondria and the Meaning of Life* (2005) and the brilliant *Life Ascending: The Ten Great Inventions of Evolution* (2009). Read them. They are wonderful.

Enormous thanks must also go to my editor at Bloomsbury, Jim Martin. Even though Jim was uneasy at times about commissioning a book about death, he stuck with it. Jim's a brilliant editor and a lovely person, which helps in every way when you're writing a difficult book like this. Thanks also to the rest of the Bloomsbury family: Laura Brooke, Anthony LaSasso, Jacqueline Johnson, Debbie Robinson, Lucy Clayton and Julie Bailey among them, along with many others too. Enormous thanks to Liz Drewitt (natureedit.com) for all of her incredibly helpful comments and edits. Also a special thank you to Anna MacDiarmid at Bloomsbury for steering the project through its final stages. My literary agent Jane Turnbull, like Jim, was never anything but supportive. Thank you.

A great deal of this book was written in Market Harborough library – such a quiet and warm place (we are enormously privileged to have libraries, please use them!). Thank you to all of the staff there, particularly Emily Warren. Thank you to Ruth Kent, my faithful and very supportive cuttings editor. And an enormous thank you too to Sam Goodlet for her fantastic chapter headers (you can find out more about Sam at www.samdrawsthings.co.uk).

The 'What's the point?' anecdote given about the young person in the epilogue is a true story (it's actually happened to me more than once) but the response I should had given didn't come to me courtesy of my own ruminations. It came to my mind because of a similar anecdote I had once heard about Carl Sagan, who apparently encouraged young people to 'choose something meaningful to do with their lives' when they asked such questions. It's a touching response (my advice: if you want to live forever, be like Carl Sagan).

Thank you to my mum and dad for all their love and never-ending support. Thank you so much for everything. You've taught me so much about life and death. Thank you. And final thanks must go to my wife, Emma, who has offered me so much love and kindness and not once wrinkled her nose when I came home (only a few times) stinking of death. It's not easy being married to a penniless writer, but somehow she puts up with it. Thank you, thank you, thank you. We found meaning in life together and every part of this life has been made wonderful through you. I love you to death … and back. x

Index